T0199253

Blockchain Technology in Supply Chain Management for Society 5.0

Society 5.0 is a human-centered community where integrated systems operate throughout society to secure comfort in all aspects of life, from energy and medical care, to education, work, and leisure. Blockchain technologies enable the streamlining of supply chain processes and information sharing among various industries. This book presents recent research on the adaptation and implementation of Blockchain technologies in supply chain management in Society 5.0. It discusses different applications of blockchain, its important role in connecting information technology and artificial intelligence with human lives, the challenges, and the future of supply chain management for societal improvements.

Smart and Intelligent Computing in Engineering

Series Editor:
Prasenjit Chatterjee, Morteza Yazdani, Dragan Pamucar, and Dilbagh Panchal

Artificial Intelligence Applications in a Pandemic COVID-19
Salah-ddine Krit, Vrijendra Singh, Mohamed Elhoseny, and Yashbir Singh

Advanced AI Techniques and Applications in Bioinformatics
Loveleen Gaur, Arun Solanki, Samuel Fosso Wamba, and Noor Zaman Jhanjhi

IoT-Based Smart Waste Management for Environmental Sustainability
Biswaranjan Acharya, Satarupa Dey, and Mohammed Zidan

Applications of Computational Intelligence in Concrete Technology
Sakshi Gupta, Parveen Sihag, Mohindra Singh, and Utku Kose

Machine Learning Algorithms and Applications in Engineering
Prasenjit Chatterjee, Morteza Yazdani, Francisco Fernández-Navarro, and Javier Pérez-Rodríguez

Blockchain Technology in Supply Chain Management for Society 5.0
K Mathiyazhagan, Atour Taghipour, and Vernika Agarwal

For more information about this series, please visit: https://www.routledge.com/Smart-and-Intelligent-Computing-in-Engineering/book-series/SICE

Blockchain Technology in Supply Chain Management for Society 5.0

Edited by
K Mathiyazhagan
Atour Taghipour
Vernika Agarwal

CRC Press
Taylor & Francis Group
Boca Raton London New York

CRC Press is an imprint of the
Taylor & Francis Group, an **informa** business

First edition published 2023
by CRC Press
6000 Broken Sound Parkway NW, Suite 300, Boca Raton, FL 33487-2742

and by CRC Press
4 Park Square, Milton Park, Abingdon, Oxon, OX14 4RN

CRC Press is an imprint of Taylor & Francis Group, LLC

ISBN: 978-1-032-00629-1 (hbk)
ISBN: 978-1-032-01165-3 (pbk)
ISBN: 978-1-003-17743-2 (ebk)

DOI: 10.1201/9781003177432

Typeset in Palatino
by codeMantra

Contents

Section 1   Introduction

Section 2   Barriers in Adoption

Section 3   Opportunities for Future

Section 4   Industry Application

Preface

The recent transformation of the information society (Society 4.0) into analytical data-driven society 5.0, the data scientist and industry practitioners are looking for novel decision-making settings in the supply chain (SC) analytics. Most of the data available right now have an overflow of information, unstructured and low in quality. The task of finding the necessary information from this overflowing information and analyzing it is a burden for supply chain analytics. To contribute towards this changing SC dynamics, the current issue focuses on the blockchain solutions for understanding the extent of implementation of society 5.0 in supply chain management. Society 5.0 is a society, where multiple industries are interconnected with focus on interface, security, big data and financial infrastructure. Society 5.0 advocates an ecosystem in which cyberspace and physical space are integrated to support an affluent and human-friendly society. Data scientists regard the interconnection of industry and society through information technologies. Blockchain is considered to be at the core of such a cyber ecosystem. Blockchain is an essential component not only of financial transactions but it can overall aid in interconnecting the entire SC. The wave of digitalization in various industries is expected to enable a business model transformation to interconnect industries and solve various social problems, as an outcome of Society 5.0.

Blockchain technologies enable streamlining of SC processes and information sharing among various industries, thereby bringing new business opportunities. Blockchain is a decentralized, distributed database that maintains a continuously growing list of secure data records. Blockchain technology will reduce time and cost while assuring safety in global business transactions. Blockchain allows multiple parties to securely interact with the same universal source of truth. It allows more transparency between supply chain entities for reducing counterfeits, authenticity of information, and reducing conflicts. In these circumstances, blockchain technology opens important avenues to make efficient and fair use of data. In a broader sense, this technology is also referred to as a "decentralized ledger," which can involve a large number of unspecified people to contribute to the effective and fair use of data in a decentralized manner. In summary, blockchain is expected to play a significant role in connecting information technology and technologies such as artificial intelligence (AI), Internet of Things (IoT), and big data with our lives.

Together an amalgamation of blockchain technologies in society 5.0 can aid the global community in building better knowledge for actions that will strengthen SC networks. This book is planned to provide a stage-by-stage integration of blockchain technologies in society 5.0 to achieve societal benefits.

The book's various chapters are expected to support the stakeholders in the implementation of blockchain technologies for making society 5.0 successful. The book chapters will be organized into four parts: (1) Introduction to blockchain and society 5.0, (2) Barriers in adoption of blockchain technologies in society 5.0, (3) Opportunities for implementation of blockchain in society 5.0, and (4) Industry applications demonstrating the blockchain in society 5.0.

Chapter 1 discusses the idea of the resilient supply chain, the barriers to the resilient supply chain, and the role of society 5.0 capabilities in enabling the resilient supply chain. The unprecedented events in the last decade have proved that the business's success in the twenty-first century has always been at risk. The uncertainties causing the business risks include cross-border relationships, global economic conditions, natural calamities, competitiveness, fast-changing customer needs, technology, etc. Among these uncertainties, the technological up-gradation in manufacturing and supply chain operations has developed immense capabilities that improve the efficiency and effectiveness of the business and enable resiliency against disruption. Thus, the concept of a resilient supply chain will aid the stakeholders in implementing the society 5.0 to a wider spread as it will aid in incorporating the technological uncertainties in the businesses.

Chapter 2 discusses the need for blockchain in reducing the issues faced by supply chain. Challenges and benefits in the field of supply chain for society 5.0 with the help of Blockchain help us to understand the content more effectively. It greatly helps in the payments and transparency among the participants in the supply chain for society 5.0. The supply chain is a complex system that comprises many components and stakeholders. Traceability, Transparency, and supply chain finance throughout the supply chain are attainable through this paper. Key features of Blockchain can help eliminate costly delays, and storage space. It can enhance transactions and automatize systems.

Chapter 3 discusses the identification of challenges that would help the policymakers to plan the path towards the adoption of successful technological adoption. The study investigates sector-wise challenges of society 5.0 in emerging economies with the help of scrapping Twitter data and performing a sentiment analysis. To validate the challenges, interviews with industry experts were conducted. The chapter tries to give implications for the policymakers, businesses, and economies to widen their scope of technology adoption and progress towards society 5.0.

Chapter 4 elaborates the public distribution system (PDS) of India and the use of blockchain in enhancing the world's largest distribution network. This book chapter studies the impediments in blockchain adoption to Indian PDS with the help of industry experts' opinion. The prioritization of identified barriers in this chapter will be useful for stakeholders and government authorities to take necessary actions according to the weight of each barrier and this is done using a Best worst Method (BWM) technique. This chapter helps us to understand how to deal with the challenges of blockchain implementation. This chapter suggests designing a proper system with adequate governance to promote the blockchain adoption design for the smooth working of the system.

Chapter 5 elaborated the need for blockchain technologies (BCT) in supply chains (SCs). Motivated by the COVID-19 outbreak and the barriers it imposes on sustainable SC, the chapter explores the driving and restraining forces related to BCT adoption in sustainable SC to tackle network disruptions. A comprehensive analysis of BCT technology, sustainable SC, and driving as well as restraining forces for espousing BCT to achieve SC with improved sustainability is discussed in this chapter. This chapter discusses a systematic study of how BCT fits in the modern SC management system and explores possible challenges with its application.

Chapter 6 elaborates the growth of the new technologies and their impact on supply chains. The chapter emphasizes the need for supply chain professionals to constantly train and reskill themselves to work more efficiently to prepare for the future. Fast progressions in digital transformation are reforming global supply chains and changing the present models of the procurement inside companies. Vastly, procurement involves the exchange of data and information between the actors of the supply chains. As the complexity increases, numerous inventions are developed to support the procurement in the form of automated solutions for different industries. These solutions enclose blockchain, data analytics, robotics, internet of goods and smart contracts, etc. Therefore, this chapter aims to explore the possibilities of these technologies in intelligent procurement and the main obstacles against their adaption.

Chapter 7 discusses opportunities and challenges for digital technologies in SCs, society 5.0 outlook, role of government in promoting humanitarian aspects of SC, and how technology laggards can adopt them, so that the world can be prepared for society 5.0. For all digital technologies, data is the key, so the process starts with Internet of Things (IoT) for data acquisition, Blockchain Technology for secure data storage and smart contracts usage, data science and machine learning to extract insight and predict the future, and lastly Artificial Intelligence to automate the knowledge for greater experience. Society 5.0 is aiming to establish economic balance by linking physical space with cyberspace, which will be an advancement for the current information age, in this initiative digital technologies play a critical role.

Chapter 8 explores the concept of the human-centred technological evolution of society 5.0, the biggest opportunities and challenges of its implementation in supply chain management, among other related topics and technologies, such as industry 4.0 and supply chain 5.0. Society 5.0 is the end phase of four prior phases, beginning with the first society of hunting, and ends with the society of information. Supply chain management can be regarded as the group of activities that focus on the good and service flow throughout the environment. Both the concepts of society 5.0 and supply chain management intertwine on numerous points, mainly in the application of society 5.0 technologies in the addressing of common supply chain problems.

Chapter 9 discusses the development of information technology, in supply chains, which has shown that this digital revolution can be a source of performance for enterprises and governments. Among these technologies is the blockchain, which is a stored database with credibility and trust of all actors in the supply chain. The actors of supply chain benefit from a secure and guaranteed information to carry out their transactions. The application of blockchains in cryptocurrency not only reduces risk of information, but also eliminates several processing and transaction fees. This allows countries with volatile currencies to have a more stable currency. The objective of this chapter is to conduct a literature review to synthesize the knowledge of using blockchains in the cryptocurrency.

Chapter 10 discusses the use of blockchain technology in various domains to enhance solutions faced by these domains. Healthcare has also emerged as one such field where this technology is being implemented to solve many day-to-day issues. Patient Data Management is one field where hospitals, laboratories, pharmacies, and physicians have started using a blockchain network. Using blockchain technology enhances the healthcare system's efficiency in terms of performance, transparency, security, and consistency. This paper elaborates two models: one for implementing Blockchain technology for Patient Data management and another, for tracing the drugs given to a patient. Both the models help in increasing the accuracy and correctness in treating a patient.

Finally, Chapter 11 focuses on the capacity building of stakeholders for the education service supply chain with the society 5.0 dimension. Under this scenario, we would like to understand that given newer technologies are coming in the education sector, one such being the blockchains, these stakeholders need to keep themselves up to date. The role of heads of institutions will play a major role in overhauling the entire education system to make it blockchain compliant. Therefore, this paper aims to understand the needs of the Heads of Institutions from the teachers in the blockchain environment. The blockchains in the education systems can only be developed if the capacity building of the system stakeholders can be done. The Quality Function Deployment (QFD) methodology is used to understand what the teachers should be able to do in a blockchain-based education institution. It then links it up with how they will be able to achieve the "whats" through various methods of training. QFD aims to let us know which of the "hows" is important and needs to be concentrated upon to develop an efficient blockchain-based learning system.

This book is aimed at both academics and practitioners alike. The potential audience included researchers, the manufacturing industry, supply chain managers, and policymakers who are focused on building smarter societies with the new age technologies. The book focuses on understanding the new age technology of blockchain and its implementation in developing society 5.0. Specifically, the book provides an essential framework for policy-makers and executives for strategic and tactical decision-making. The book can be used as a textbook or supplementary task for the graduate program in supply chain management, operations management, smart cities, sustainable management, and business administration. Society 5.0 is an emerging and critical subject that has not been comprehensively addressed in the literature.

Acknowledgements

This book comprises the collective work of many individuals from a wide range of organizations. It would not have been possible for us to write this book without the contributions of all these authors. Firstly, we would like to appreciate the support and cooperation given by Thiagarajar School of Management, Madurai, Tamilnadu, India, Normandy University, Le Havre, France and Amity University, Noida, Uttar Pradesh, India during the various stages of this work. In addition, we would also like to that the staff at CRC Press for their enthusiasm in supporting our endeavor. They encouraged us from the beginning and throughout the publication process. Finally, we would like to thank our families for all of their support during the writing and editing of this book. It was the constant faith of our families that encouraged us to complete this work.

Editors

K Mathiyazhagan, PhD, is Chairperson of Research Centre and Associate Professor at Thaigarajar School of Management, Madurai, Tamilnadu. He is an associate editor in Management Decision (Emerald) impact factor 4.9 and *Environment, Development and Sustainability Journal* (Springer) with an impact factor of 3.2 and has more than 3,000 research citations. He has edited special issues in *Journal of Business Logistics* (JBL), *International Journal of Physical Distribution and Logistics Management* (IJPDLM), *International Journal of Logistics Management* (IJLM), *Sustainable Production and Consumption* (Elsevier), *Socio-Economic Planning Science* (Elsevier). His publications are in IJPE, IJPR, PPC, IEEE TEM, JCP, RCR, IJAMT, etc. Also, he is an editorial member in many reputed journals. One of his papers was awarded as Excellence Citation Award by Emerald Publisher Ltd. In addition, he has edited international books on Blockchain and Lean-Green and Society 4.0 with Elsevier, Taylor & Francis, and Springer. He is a visiting faculty at the University of Rome Tor Vergata, Italy. Dr. Mathiyazhagan is an active reviewer of more than 30 reputed international journals and organized two international conferences and workshops. He has received best reviewer certificates from reputed journals. He has supervised more than ten undergraduate and three postgraduate and three PhD scholars. Dr. Mathiyazhagan has edited many international conference proceedings that have been published by Springer. His research interest is green supply chain management; sustainable supply chain management; multi-criteria decision making; third-party logistic provider; sustainable lean manufacturing, public distribution system; and Lean Six Sigma. He has more than 10 years of research and teaching experience.

Atour Taghipour, PhD, is a Professor of Operations Management and the Head of an international management master program at Normandy University, France (Master (II) of International Purchasing). He holds an HDR in management from Normandy University and a PhD in industrial engineering from the Polytechnic School of Montreal, Canada. He earned two master's degrees, one in management, logistics and strategy and other in industrial engineering. He has more than 10 years of experience as a manager in the automobile industry. He has published several books and many research papers in international journals. His areas of research are supply chain and operations management.

Vernika Agarwal, PhD, is an Assistant Professor in operations management at Amity International Business School at Amity University, Noida, Uttar Pradesh. She works in the fields of sustainable supply chain management, multi-criteria decision making, third-party logistic providers, sustainable lean manufacturing, circular economy and sustainability, cross-disciplinary research in supply and operations management, optimization, reverse logistics, and empirical research. She earned a PhD in operations and supply chain management in the Department of Operational Research at the University of Delhi, India. She has published or presented several papers in reputed national and international journals.

Contributors

Basma Addakiri
Faculty of International Business
Normandy University
France

Vernika Agarwal
Amity International Business School
Amity University
Noida, Uttar Pradesh, India

Yazeed Abdul-Monem Al-Mahi
Department of Industrial
   Engineering
Faculty of Engineering
King Abdulaziz University
Jeddah, Saudi Arabia

Salman Ali Al-oufi
Department of Industrial
   Engineering
Faculty of Engineering
King Abdulaziz University
Jeddah, Saudi Arabia

R. Anand Babu
Department of Information
   Technology
E.G.S. Pillay Engineering College
Nagapattinam, Tamilnadu, India

Abdul Zubar Hameed
Department of Industrial
   Engineering
Faculty of Engineering
King Abdulaziz University
Jeddah, Saudi Arabia

Mukund Janardhanan
School of Engineering
University of Leicester
Leicester, United Kingdom

P. Kalpana
Department of Mechanical
   Engineering
Indian Institute of Information
   Technology Design and
   Manufacturing (IIITDM)
Kancheepuram, Tamilnadu, India

R. Karthi
Department of Master of Business
   Administration
E.G.S. Pillay Engineering College
Nagapattinam, Tamilnadu, India

Mohammad Hani Kashif
Department of Industrial
   Engineering
Faculty of Engineering
King Abdulaziz University
Jeddah, Saudi Arabia

Arshia Kaul
Anil Surendra Modi School of
   Commerce
NMIMS University
Mumbai, India

Vasundhara Kaul
Carpediem EdPsych Consultancy
   LLP
Mumbai, India

Sharad Khattar
Amity International Business School
Noida, Amity University
Uttar Pradesh, India

Yadhukrishnan Vyppukkaran Krishnan
School of Engineering
University of Leicester
Leicester, United Kingdom

R. Lavanya
Department of Information
   Technology
E.G.S. Pillay Engineering College
Nagapattinam, Tamilnadu, India

K. Lenin
Department of Mechanical
   Engineering
K. Ramakrishna College of
   Engineering
Tirchurappalli, Tamilnadu, India

Leninisha S
School of Computer Science and
   Engineering
Vellore Institute of Technology
Chennai, Tamilnadu, India

XiaoWen Lu
Faculty of International Business
Normandy University
France

Snigdha Malhotra
Amity International Business School
Amity University
Noida, Uttar Pradesh, India

Marina Marinelli
School of Engineering
University of Leicester
Leicester, United Kingdom

K. Nagalakshmi
Department of Information
   Technology
E.G.S. Pillay Engineering College
Nagapattinam, Tamilnadu, India

Nivedha M
School of Computer Science and
   Engineering
Vellore Institute of Technology
Chennai, Tamilnadu, India

Parkavi K
School of Computer Science and
   Engineering
Vellore Institute of Technology
Chennai, Tamilnadu, India

Bharat Singh Patel
Thiagarajar School of Management
Madurai, Tamilnadu, India

K. Raju
Department of Information
   Technology
E.G.S. Pillay Engineering College
Nagapattinam, Tamilnadu, India

S. Ramabalan
Department of Mechanical
   Engineering
E.G.S. Pillay Engineering College
Nagapattinam, Tamilnadu, India

Deepali Ratra
Jagan Institute of Management
   Studies (JIMS Rohini)
New Delhi, India

K. Raj Kumar Reddy
Data Scientist
Chennai, Tamilnadu, India

Seema Sahai
Amity International Business School
Amity University
Noida, Uttar Pradesh, India

Cherian Samuel
Department of Mechanical
    Engineering
Indian Institute of Technology
    (BHU) Varanasi
Varanasi, India

Mustafa Smahi
Faculty of International Business
Normandy University
France

Aishwariya Subakkar
School of Computer Science and
    Engineering
Vellore Institute of Technology
Chennai, Tamilnadu, India

Atour Taghipour
Faculty of International Business
Normandy University
France

Neelesh Thallam
Amity International Business School
Amity University
Noida, Uttar Pradesh, India

J. Vanitha
Department of MCA
E.G.S. Pillay Engineering College
    (Autonomous)
Nagapattinam, Tamilnadu, India

Ajeet Kumar Yadav
Department of Mechanical
    Engineering
Indian Institute of Technology
    (BHU) Varanasi
Varanasi, India

Section 1

Introduction

1

A Society 5.0 Approach for Developing a Resilient Supply Chain

Ajeet Kumar Yadav, Bharat Singh Patel, and Cherian Samuel

CONTENTS

1.1 Introduction

The situation arising out of the outbreak of Corona Virus Disease (COVID-19) is unprecedented, frightening, widespread, and has a global impact (Baral et al., 2022). It took the lives and livelihoods of millions of people across the globe and impacted the world economy badly. It has also challenged human preparedness to deal with uncertainties and their ability to counter the disruptions (Patel and Sambasivan, 2022). The world has witnessed many more similar events in the past, such as the Tsunami (2004), the Japan Earthquake (2011), the 9/11 terrorist attack (2001), fire at the Phillips semiconductor plants (2000), etc. Along with COVID-19, the most recent incident that shook world business is the Suez Canal crisis, causing almost 12% of the world freight blockage. The need to address these unprecedented events is because of the global business connectedness in the Supply Chain (SC). Therefore, disruption at any point of the SC affects the global business scenario irrespective of the region of occurrence.

Considering the impact of the SC disruptions, the term Resilient Supply Chain (RSC) comes into the picture, consisting of proactive and reactive strategies to mitigate and recover from the SC-related disruptions. Christopher and Peck (2004) defined RSC as the ability to mitigate disruptions and regain its initial or even better position post disruptions. Kumar and Anbanandam (2020) defined RSC as responding quickly to unforeseen disruptions and gaining a competitive advantage. Researchers have attributed the RSC concept to various abilities such as anticipative, resistive, absorbing, springing back, preservance, development, growth, learning, adaptability, etc. To achieve these capabilities in the SC, experts have proposed several practices, such as SC flexibility (Singh Patel et al., 2020), SC agility (Patel et al., 2018), SC collaboration (Radhakrishnan et al., 2018), SC sensitivity (Rocío Ruiz-Benítez et al., 2018), SC restructuring (Duchek, 2020), virtual enterprising (Patel et al., 2020), etc. Moreover, Hohenstein et al. (2015) focused on identifying the possible enablers of the RSC. Yadav and Samuel (2022) focused on identifying the most influencing while considering the causal-dependence relations among them, whereas Jain et al. (2017) intended to quantify the degree of influence one resilience practice is having over the other. Furthermore, Agarwal and Seth (2021) focused on quantifying the resilience score of the SC using the graph theory approach.

The recent disruptions have proven that whatever steps we take in enabling resilient behavior, the impact of disruptions is unavoidable. These failures indicate the presence of inhibitors that prohibits the successful implementation of RSC strategies (Agrawal and Seth, 2021). Also, as per the force field theory, the existence of a system pertains only when force supports the system more than the force opposing the system. Thus, the elimination of these barriers needs to be considered an integral part of the resilient practices of the SC. Therefore, identifying the obstacles to the RSC is of utmost need. Some of the most influencing barriers are poor trust among SC members, poor SC structure, poor cooperation and collaboration during adversities, etc.

Since the evolution of Society 5.0 concept, the visibility and the management of the SC and related operations have been eased a lot. Society 5.0 is considered to be a business-wide concept that interlinks all the SC and related activities on a digital platform (Nayernia et al., 2021). It includes real-time access and the control of the production system and logistics services; along with data essential for the business process autonomously (Ralston and Blackhurst, 2020). The concept of Society 5.0 was coined by German government officials in 2011 at Hannover Messe, suggesting the development of an intelligent system across the SC that integrates humans and machines to adapt and grow efficiently with the changing customer needs and the ecological conditions

(Kuo and Kusiak, 2019). According to Frederico (2021), Society 5.0 technologies improve business performance during normal business operations and minimize disruptions and their impact by enabling resilient capabilities and mitigating the inhibitors of the RSC. In this chapter, we carried out the following objectives:

1. What are RSC and its associated capabilities?
2. Which of the factors is enabling and obstructing the success of the RSC?
3. How do Society 5.0 technologies profound the RSC?

1.2 Literature Review

1.2.1 Resilient Supply Chain

The resilience concept of the SC is derived from various disciplines such as engineering, ecological characteristics, human behavior, community stabilization, etc. (Annarelli et al., 2020). Therefore, the resilience concept of the SC is characterized as a multi-phenomenon concept that includes forecasting, preparation against the undesired situation, learning and growing with the changing conditions, etc. It also includes opportunity capitalization, recovery post disruptions, resisting unexpected conditions, and adapting to the changes to achieve business goals (Duchek, 2020). Furthermore, Aslam et al. (2020) gave importance to the recovery ability of the SC, whereas Kumar and Anbanandam (2020) gave importance to the preparation against vulnerabilities, adequate response to uncertain events, and efficient and effective recovery against disturbances. This capability improves the competitiveness and effectiveness of the SC.

Moreover, the authors like Sangari and Dashtpeyma (2019) and Parast et al. (2019) focused on reactive strategies that maintain the SC performance post disruptions. Resilience ability of the SC comprises the ability to adapt to the variability in supply and demand, quickness in decision-making and production processes, efficient SC operations, capability improvement, and post-disruption development (Tukamuhabwa et al., 2015). Ajeet and Samuel (2021) classified the RSC capabilities into proactive and reactive thereby providing a more elaborated set of capabilities: anticipation, preparedness, responsiveness, recovery, adaptive, resistive, learning, and innovation. Hollnagel et al. (2006) specifically focused on improving the ability of the SC to anticipate, detect, and defend vulnerabilities.

Thus, from the above discussion, we conclude that a RSC is the consequence of the resilient practices of the SCs and is defined by the characteristics as shown in Figure 1.1. Hence, it can be concluded that the resilient capabilities

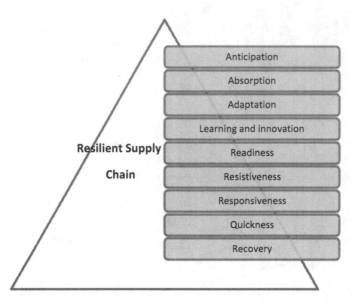

FIGURE 1.1
Characteristics of the RSC.

of the SCs not only mitigate the impacts of the disruptions but also enable sustainability in the supply chain (Jain et al., 2017). The various practices that enable the resilient ability of the supply chain have been given different names by the researchers, such as strategies, antecedents, factors, attributes, principles, elements, etc. Based on their ability, these practices are clubbed into different groups, such as proactive and reactive practices (Lohmer et al., 2020), practices for improving the anticipation ability, practices for absorbing the disruption impacts, and practices for recovering from the disruptions (Ali et al., 2017), etc. Table 1.1 provides the detailed RSC practices present in the literature.

1.2.2 Obstruction to the RSC

Since the development of the resilient concept, many organizations tend to develop the resilient capability of their organizations and the associated organizations. However, the socioeconomic and ecological vulnerabilities of the twenty-first century have proven that howsoever we strategize the resilient capability of the organizations and their SC; they cannot withhold their performance during disruptions. Thus, suggesting the presence of inhibitors prevents the successful implementation of resilient capabilities (Ali et al., 2017). Therefore to enable the true resilient capability to the SC, it is essential to identify and eradicate these inhibitors (Rajesh, 2017). Also, this phenomenon is in accordance with the force field theory, which suggests that the existence of a system is only possible when the driving force is equal or greater than the barriers or opposing force. However, the identification and evaluation of the barriers to the success of the SC is not a very new concept and has been used in improving SC efficiency, sustainability, and growth. However, identification and evaluation of the RSC's barriers have gained very little importance compared to developing the resilient capability of the SC (Pereira et al., 2014). Some of the essential barriers are provided in Table 1.2.

1.3 Society 5.0

The conceptual evolution of the term Society 5.0 is the outcome of technological improvement and the digitization of the production and SC practices. As per the cabinet office, Society 5.0 is *"A human-centric Society that promotes the economic development with the focus of minimizing societal problems by a system that comprises an integrated cyberspace and physical space."* This concept was proposed in the 5th Science and Technology Basic Plan for socioeconomic development of Japan. It is the result of the continuous societal, economic, and industrial development across the globe. Moreover, it is assumed that Society 5.0 concept followed Society 1.0 (hunting society), Society 2.0 (agricultural society), Society 3.0 (power generation), and Society 4.0 (Information Society). The technical and digital advancement in SC and related operations enables real-time communication among various SC nodes (Ralston and Blackhurst, 2021). Society 5.0 allows autonomous decision-making at different SC stages to mitigate vulnerabilities, gain opportunities, and improve overall productivity.

Moreover, Nayernia et al. (2021) suggested that the successful implementation of the technological capacities brings a high level of control over the information and its quality and improves re-configurability through decentralized decision-making. Moreover, Zhang et al. (2020) and Ivanov et al. (2019) suggested that technological advancement enables risk identification and avoidance through proactive decision-making, thus improving the business continuity and effectiveness of the SC. According to Ralston and Blackhurst (2020), the industry 4.0 capabilities embark Society 5.0 that results in an intelligent SC, i.e., better, faster, accurate, and reliable, with critical thinking and adaptive capacity.

Soares et al. (2021) gave importance to machine-to-machine communication, integration of robotics, consistent communication, innovative drive technologies, and real-time tracking and tracing. Thus, the successful implementation of Society 5.0 will enable the resilient capability to the SC. It allows the development of resilient characteristics to the supply through the successful implementation of resilient practices. Moreover, Chofreh et al. (2020) explained the benefits of industrial advancement in improving the trust among the SC members, collaborating planning and decision-making,

TABLE 1.1

Important Literature Related to RSC Practices

Literature RSC Practices/Factors	Soni et al. (2014)	Hohenstein et al. (2015)	Kamalahmadi and Parast (2016)	Azevedo (2016)	Ali et al. (2017)	Jain et al. (2017)	Rajesh (2017)
Adaptability	✓					✓	
Agility	✓		✓	✓		✓	✓
Anticipation					✓		
Assortment planning							
Capacity development							
Capacity enhancement							✓
Capacity flexibility							
Collaboration	✓	✓	✓		✓	✓	✓
Cooperation							
Cross training							
Contractual agreement							
Continuity					✓		
Dispersion							
Demand centralization							
Efficiency							
Financial strength							
Flexibility		✓		✓	✓		✓
Flexible transportation				✓			
Human resource management		✓					
Information sharing			✓		✓	✓	
Inventory management		✓					✓
Innovation			✓		✓		
Insurance							
Knowledge management					✓		
Knowledge sharing	✓						
Lead time reduction				✓			
Leadership			✓				
Manufacturing flexibility							
Market position security							
Market sensitivity						✓	
Multiple sourcing				✓			
Planning capability							✓
Pricing capability							✓
Postponement							✓
Product rollover capability							✓
Recovery							
Redesigning					✓		✓
Redundancy		✓			✓		

Radhakrishnan et al. (2018)	Neubauer (2018)	Kochan and Nowicki (2018)	Hosseini et al. (2019)	Sabahi and Parast (2020)	Lohmer et al. (2020)	Rajesh (2020)	Yadav and Samuel (2021)	Agarwal et al. (2021)
✓		✓					✓	
	✓	✓	✓	✓		✓	✓	✓
✓								
						✓	✓	
✓								
						✓		
✓	✓	✓	✓	✓			✓	
					✓			
						✓		
					✓			
✓								
						✓		
✓		✓						
✓		✓					✓	
✓	✓	✓	✓	✓			✓	
	✓		✓		✓		✓	
							✓	
	✓							
				✓				
	✓			✓				
						✓		
✓		✓						
							✓	
						✓	✓	
						✓	✓	
		✓						
	✓	✓		✓			✓	

(Continued)

TABLE 1.1 (*Continued*)

Important Literature Related to RSC Practices

Literature RSC Practices/Factors	Soni et al. (2014)	Hohenstein et al. (2015)	Kamalahmadi and Parast (2016)	Azevedo (2016)	Ali et al. (2017)	Jain et al. (2017)	Rajesh (2017)
Re-engineering				✓			
Restructuring	✓					✓	
Revenue management							
Risk and reward sharing	✓					✓	
Risk hedging							
Culture of risk management	✓		✓		✓	✓	
Robustness					✓		
Security					✓		
Standardization							✓
Strategic stocking				✓			
Supply contract flexibility							
SC design							
Supply base strengthening							
Sustainability	✓					✓	
Technology capability						✓	
Trust	✓		✓			✓	
Velocity		✓	✓				
Visibility	✓	✓	✓	✓	✓	✓	

visibility of the network, etc., thus leading to sustainability. This concept improves the SC capabilities at each SC node, such as planning, execution, and control. Spieske and Birkel (2021) proclaimed that digitization enables self-configuration, self-adjustments, and self-optimization against vulnerabilities and disturbances. Henceforth, it can be said the implementation of Society 5.0 concept will allow the resilient capability to the supply chains through the successful implementation of the resilient strategies, minimization or eradication of the barriers, and developing the intelligent systems for production, distribution, and control. Considering the commonality and the capabilities of Society 5.0 technologies discussed in the previous studies, they can be broadly described into the following seven technologies, as discussed below.

1.3.1 Cloud Computing

It is an alternative to rigid internal infrastructure for real-time decision-making through data management (Nayernia et al., 2021) and enables the successful implementation of the Internet of things (IoT) by integrating soft resources (Urban et al., 2020). Parast et al., (2019) suggested that the company can

Radhakrishnan et al. (2018)	Neubauer (2018)	Kochan and Nowicki (2018)	Hosseini et al. (2019)	Sabahi and Parast (2020)	Lohmer et al. (2020)	Rajesh (2020)	Yadav and Samuel (2021)	Agarwal et al. (2021)
	✓							✓
							✓	
		✓					✓	
						✓		
✓				✓			✓	✓
						✓		
							✓	
						✓		
					✓	✓		
						✓		
	✓			✓				
	✓	✓		✓				
✓	✓		✓	✓		✓	✓	

effectively use it to make quick decisions, thus minimizing the losses due to possible disruptions. In the presence of big data analytics (BDA), it enables the SC real-time connectivity and traceability. It allows quick collection, analysis, and interpretation of the data related to the risk, disruption, performance, etc., to make a more efficient supply, demand, and transportation planning (Arsovski et al., 2017; Ramirez-Peña et al., 2020).

1.3.2 Internet of Things

It is defined as the network of physical systems equipped with digital platforms to independently interact with the other systems across the boundary of the company (Queiroz et al., 2021). It enhanced the communication between humans and machines (Wilkesmann and Wilkesman, 2018). In the presence of data analytics, it amplifies the SC data capabilities through the development of the protocols for various devices and forecasting accuracy (Oncioiu et al., 2019). The components, tools, or technologies of the IoT include radio frequency identifiers (RFID), microphones, wireless sensors, a global positioning system (GPS), barcodes, etc., which enables the close connectivity and data collection among the various systems involved in the SC (Dunke et al.,

TABLE 1.2

Barriers to the Resilient Supply Chain

Reference	Barriers to the Resilient Supply	Centralization of Assets	Opportunistic Behavior of the Supply Chain Members	Un-Insured Critical and Risky Assets	Poor Supply Chain Network	Continuous Strive for Lean Process	Lack of Flexibility	Single Dedicated Sourcing	Poor Upstream and Downstream Performance	Operating at Full Capacity
Yadav and Samuel (2021)		✓		✓	✓	✓				
Agarwal and Seth (2021)					✓					
Khullar et al. (2020)				✓						
Singh et al. (2019)						✓			✓	
Scholten et al. (2019)							✓			
Rajesh (2020)		✓		✓	✓	✓	✓			
Hsu et al. (2021)			✓							
Xu et al. (2018)										
Ali et al. (2017)					✓				✓	
Shibin et al. (2016)										
Gorane and Kant (2015)					✓					
Vilko et al. (2014)								✓		
Yinan et al. (2014)		✓			✓				✓	
Pereira et al. (2014)										
Lavastre et al. (2014)				✓						
Cantor et al. (2014)					✓					
Samvedi et al. (2013)						✓				
Pettit et al. (2013)										
Govindan et al. (2013)		✓								
Simangunsong et al. (2012)								✓		
Erikkson and McConnell (2011)				✓						✓
Dowty and Wallace (2010)										
Blackhurst et al. (2005)			✓							
Giunipero and Eltantawy (2004)				✓					✓	✓
Lämsä and Savolainen (2000)										

Inadequate Response During the Disruptions	Poor Information	Lack of Risk Awareness	Excessive Outsourcing and Vertical Integration	Lack of Managerial Commitment	Poor Judgment and Inaccurate Forecast	Poor Financial Health	Poor Contingency Planning	Mistrust Among the Supply Chain Partners	Lack of Multitasking Workforce	Opposition to Change and Innovation Chain
		✓		✓		✓	✓	✓		✓
	✓							✓		
						✓				
✓				✓						✓
			✓							
	✓		✓		✓			✓		
				✓						
✓						✓				
		✓					✓		✓	
						✓				
		✓								
✓								✓		
			✓							
		✓								
		✓					✓			
					✓					
							✓			
✓										
				✓				✓		

2018). It allows the tracking of the SC process and the data so collected can be used as input parameters for BDA, artificial intelligence, cyber-physical system, etc. (Queiroz et al., 2021).

1.3.3 Big Data Analytics

It is the process that integrates data (structured or unstructured) available from various sources to dig out the information useful for decision-making (Spieske and Birkel, 2021). According to Papadopoulos et al. (2017), it helps in product traceability, understanding consumer behavior, inventory management, and managing the relationship between customers and suppliers. Moreover, it also enables the risk prediction, planning, adoption, and effective execution of the disruption strategies and development of the digital SC twins through simulation techniques (Ivanov and Dolgui, 2020). This leads to identifying and assessing the available opportunities and risk mitigation strategies.

1.3.4 Artificial Intelligence

It is the broad technological concept that enables adaptive decision-making and process development based on past data (Spieske and Birkel, 2021). It constitutes machine learning, mathematical and agent-based modeling, and a network-based approach (Baryannis et al., 2019) and enables autonomous decision-making during adversities and unknown or new business environments (Baryannis et al., 2019).

1.3.5 Cyber-Physical Systems

It is the technological aspect of the management that integrates the physical infrastructure into the system that is capable of managing and communicating the operations and the information related to the SC (Chen et al., 2019). It enables automation of the operations while improving monitoring and control. According to Ramirez-Peña et al. (2020), autonomous vehicles and robots play a vital role in cyber-physical systems that not only perform human jobs but also minimize the risk of human errors. For example, automated system used during COVID-19 for detecting and carrying the testing samples (Ralston and Blackhurst, 2020).

1.3.6 Additive Manufacturing

It is the opposite of the traditional machining process, where the product is developed by adding the materials in layers instead of removing material from the solid block. 3D printers are the most viable example of additive manufacturing which enables the production of parts, components, modules, etc. (Ivanov et al., 2019). It also minimizes the material's wastage and improves the responsiveness of the production systems.

1.3.7 Blockchain

According to Spieske and Birkal (2021), it is a technological capability originally used in financial engineering that decentralizes SC systems and leads to open and cryptographic peer-to-peer networks. In this SC, members are joined as the information ledger that creates stores and accesses the information in the form of digital blocks. It improves visibility, communication, and coordination, thus improving the trust among the SC members (Choi et al., 2020). According to Min (2019) and Kshetri (2018), it maintains and validates the transactions' records and identifies the risk, vulnerabilities, and available mitigating strategies.

1.3.8 Augmented Reality, Virtual Reality, and Mixed Reality (MR)

It enables the real-time interaction of the physical and virtual system and presents the images in three dimensions. It enhances control over the process, improves flexibility, and develops quick learning capabilities. For example, according to Keogh et al. (2020), the use of augmented reality (AR) headsets in the food and beverages SC provides all the information related to the production process and the position of processing plants to the SC managers. Other AR examples include smart glasses, video training, etc. Similarly, the concept of virtual reality (VR) is used as an integrated part of Society 5.0, which virtually creates the feeling of the real environment using simulation technology (Desai et al., 2014). Moreover, combining AR and VR improves the outcome and experience with digital and real objects. It is a complex and multilayered rendering tool for virtual objects.

1.4 Discussion

Developing the resilient characteristics into the SC improves the business performance and provides a competitive edge in the present vulnerable business environment. These characteristics are dependent on the network-wide successful implementation of resilient practices (Table 1.1). Though irrespective of the organizations' effort in implementing resilient practices, businesses often fail because of SC disruptions. These disruptions suggest the obstruction, which not only mitigates the effectiveness of the resilient practices but also makes the SC vulnerable. However, the conceptual development of Society 5.0 appears to be the most effective tool in managing resilient practices along with mitigating the impacts of the barriers. It improves the control over the SC through visibility, assessment, risk planning, cause-and-effect analysis, rapidity and effectiveness of decision-making, etc. Hence, implementing Society 5.0 capabilities with the resilience practices will lead to the development of the true resilience SC.

1.5 Conclusion

From the above discussion, it is evident that the business's success depends widely upon its ability to combat vulnerabilities and disruptions. This ability is called the resilient ability of the supply. Moreover, the successful implementation of these resilient capabilities in the SC is hindered by some inherent and socioeconomic activities associated with the SC. Furthermore, Society 5.0 concept is very prevalent in today's business environment to enable sustainable growth and improved productivity; however, very little importance is given to analyzing its role in enabling the resilient capability of the SC. On the other hand, the theory provided in this chapter offers a nominal understanding of the SC's resilience concept and the barriers hindering its successful implementation. Moreover, capabilities or techniques of Society 5.0, which are discussed in this chapter, provide essential support for their role in enabling the resilience capability to the SC by allowing the effective application of the resilient practices and minimizing and eradicating the barriers associated with it.

Though this chapter delivers a brief conceptual understanding of the RSC, RSC barriers, and the role of Society 5.0 technologies in enabling the resilient characteristics, it provides a considerable contribution toward the development of a truly RSC. Because it integrates all the factors that support and hinder the successful execution of the RSC strategies, this concept needs to be elaborated by explicitly identifying and defining the relationships among the resilient practices, barriers, and Society 5.0 technologies. Moreover, hindrances and their mitigating strategies to Society 5.0 can also be identified to make the RSC a reality.

1.6 Limitations and Future Scope

This chapter provides a brief understanding of the RSC, the practices enabling the RSC, the barriers obstructing the RSC, and the role of Society 5.0 in enabling true resilience characteristics to a SC. However, this study needs to be elaborated by estimating the supporting and obstructing role of the resilient practices and the barriers to the creation of the RSC, respectively.

Moreover, the success rate of the RSC in the presence of Society 5.0 technologies needs to be evaluated in varying business environments and during specific breakdowns.

References

Agarwal, N., and Seth, N. (2021), "Analysis of supply chain resilience barriers in Indian automotive company using total interpretive structural modelling", *Journal of Advances in Management Research.* https://doi.org/10.1108/JAMR-08-2020-0190

Agarwal, N., Seth, N., and Agarwal, A. (2021), "Evaluation of supply chain resilience index: A graph theory based approach", *Benchmarking: An International Journal*, Vol. 29, No. 3, pp. 735–766. https://doi.org/10.1108/BIJ-09-2020-0507

Ali, A., Mahfouz, A., and Arisha, A. (2017), "Analysing supply chain resilience: Integrating the constructs in a concept mapping framework via a systematic literature review", *Supply Chain Management*, Vol. 22, No. 1, pp. 16–39. https://doi.org/10.1108/SCM-06-2016-0197

Annarelli, A., Battistella, C., and Nonino, F. (2020), "A framework to evaluate the effects of organizational resilience on service quality", *Sustainability (Switzerland)*, Vol. 12, No. 3, pp. 1–15. https://doi.org/10.3390/su12030958

Arsovski, S., Arsovski, Z., Stefanović, M., Tadić, D., and Aleksić, A. (2017), "Organisational resilience in a cloud-based enterprise in a supply chain: A challenge for innovative SMEs", *International Journal of Computer Integrated Manufacturing*, Vol. 30, No. 4–5, pp. 409–419.

Aslam, H., Khan, A. Q., Rashid, K., and Rehman, S. U. (2020), "Achieving supply chain resilience: The role of supply chain ambidexterity and supply chain agility", *Journal of Manufacturing Technology Management*, Vol. 31, No. 6, pp. 1185–1204. https://doi.org/10.1108/JMTM-07-2019-0263.

Azevedo, S. G., Carvalho, H., and Cruz-Machado, V. (2016), "LARG index: A benchmarking tool for improving the leanness, agility, resilience and greenness of the automotive supply chain", *Benchmarking: An International Journal*, Vol. 23 No. 6, pp. 1472–1499.

Baral, M. M., Mukherjee, S., Nagariya, R., Singh Patel, B., Pathak, A., and Chittipaka, V. (2022), "Analysis of factors impacting firm performance of MSMEs: Lessons learnt from COVID-19", *Benchmarking: An International Journal.* https://doi.org/10.1108/BIJ-11-2021-0660

Baryannis, G., Validi, S., Dani, S., and Antoniou, G. (2019), "Supply chain risk management and artificial intelligence: State of the art and future research directions", *International Journal of Production Research*, Vol. 57, No. 7, pp. 2179–2202.

Blackhurst, J., Craighead, C.W., Elkins, D. and Hangfield, R.B. (2005), "An empirically derived agenda of critical research issues for managing supply-chain disruptions", *International Journal of Production Research*, Vol. 43, No. 19, pp. 4067–4081.

Cantor, E. D., Blackhurst, J., Pan M,., and Crum, M. (2014), "Examining the role of stakeholder pressure and knowledge management on supply chain risk and demand responsiveness", *The International Journal of Logistics Management*, Vol. 25 No. 1, pp. 202–223.

Chen, H. Y., Das, A., and Ivanov, D. (2019), "Building resilience and managing post-disruption supply chain recovery: Lessons from the information and communication technology industry", *International Journal of Information Management*, Vol. 49, No. 1, pp. 330–342.

Chofreh, A. G., Goni, F. A., Klemeš, J. J., Malik, M. N., and Khan, H. H. (2020), "Development of guidelines for the implementation of sustainable enterprise resource planning systems", *Journal of Cleaner Production*, Vol. 244, No. 20, pp. 118655. https://doi.org/10.1016/j.jclepro.2019.118655

Choi, T., Rogers, D., and Vakil, B. (2020), "Coronavirus is a wake-up call for supply chain management", *Harvard Business Review*, Vol. 27, pp. 36. https://hbr.org/2020/03/coronavirus-is-a-wake-up-call-for-supply-chain-management

Christopher, M., and Peck, H. (2004), "Building the RSC", *International Journal of Logistics Management*, Vol. 15, No. 2, pp. 1–14.

Desai, P. R., Desai, P. N., Ajmera, K. D., and Mehta, K. (2014), *A review paper on oculus rift-A virtual reality headset*. ArXiv, 1408.1173 [Cs]. Retrieved from http://arxiv.org/abs/1408.1173.

Dowty, R. A., and Wallace, W. A. (2010), "Implications of organizational culture for supply chain disruption and restoration", *International Journal of Production Economics*, Vol. 126 No. 1, pp. 57–65.

Duchek, S. (2020), "Organizational resilience: A capability-based conceptualization", *Business Research*, Vol. 13, pp. 215–246. https://doi.org/10.1007/s40685-019-0085-7

Dunke, F., Heckmann, I., Nickel, S., and Saldanha-da-Gama, F. (2018), "Time traps in supply chains: Is optimal still good enough?" *European Journal of Operational Research*, Vol. 264, No. 3, pp. 813–829.

Eriksson, K. and Mcconell, A. (2011), "Contingency planning for crisis management: Recipe for success or political fantasy?", *Policy and Society*, Vol. 30 No. 2, pp. 89–99.

Frederico, G. F. (2021), "Supply Chain 4.0 to Supply Chain 5.0: Findings from a systematic literature review and research directions", *Logistics*, Vol. 5, No. 49, pp. 1–21. https://doi.org/10.3390/logistics5030049

Giunipero, L. C. and AlyEltantawy, R. (2004), "Securing the upstream supply chain: a risk management approach", *International Journal of Physical Distribution and Logistics Management*, Vol. 34, No. 9, pp. 698–713.

Gorane, S. J. and Kant, R. (2015), "Modelling the SCM implementation barriers", *Journal of Modelling in Management*, Vol. 10, No. 2, pp. 158–178.

Govindan, K., Popiuc, M. N., and Diabat, A. (2013), "Overview of coordination contracts within forward and reverse supply chains", *Journal of Cleaner Production*, Vol. 47, No. 1, pp. 319–334.

Hohenstein, N., Feisel, W., and Hartmann, E. (2015), "Research on the phenomenon of SC resilience", *International Journal of Physical Distribution and Logistics Management*, Vol. 45, No. ½, pp. 90–117.

Hollnagel, E., Woods, D., and Leveson, N. (2006), "Resilience engineering: Concepts and precepts", *Quality and Safety in Healthcare*, Vol. 15, No. 6, pp. 447–448.

Hosseini, S., Ivanov, D., and Dolgui, A. (2019), "Review of quantitative methods for supply chain resilience analysis", *Transportation Research Part E: Logistics and Transportation Review*, Vol. 125, pp. 285–307. https://doi.org/10.1016/j.tre.2019.03.001.

Hsu, C.H., Chang, A.Y., Zhang, T.Y., Lin, W.D. and Liu, W.L. (2021), "Deploying resilience enablers to mitigate risks in sustainable fashion SCs", *Sustainability*, Vol. 13 No. 5, p. 2943.

Ivanov, D., and Dolgui, A. (2020), "A digital supply chain twin for managing the disruption risks and resilience in the era of Industry 4.0", *Production Planning and Control*, Vol. 32, No. 9, pp. 775–788.

Ivanov, D., Dolgui, A., and Sokolov, B.. (2019), "The impact of digital technology and Industry 4.0 on the ripple effect and supply chain risk analytics", *International Journal of Production Research*, Vol. 57, No. 3, pp. 829–846.

Jain, V., Kumar, S., Soni, U., and Chandra, C. (2017), "Supply chain resilience: Model development and empirical analysis", *International Journal of Production Research*, Vol. 55, No. 22, pp. 6779–6800.

Kamalahmadi, M., and Parast, M. M. (2016), "A review of the literature on the principles of enterprise and SC resilience: Major findings and directions for future research", *International Journal of Production Economics*, Vol. 171, No. 1, pp. 116–133.

Keogh, J. G., Dube, L., Rajeb, A., Hand, K. J., Khan, N., and Dean, K. (2020), *Building the future of food safety technology – The future food chain: Digitization as an enabler of Society 5.0*, pp. 11–38. https://doi.org/10.1016/B978-0-12-818956-6.00002-6.

Khullar D, Bond A. M., and Schpero W. L. (2020), "COVID-19 and the financial health of US hospitals", *JAMA Health Forum*, Vol. 323, No. 21, pp. 2127–2128.

Kochan, C. G., and Nowicki, D. R. (2018), "Supply chain resilience: A systematic literature review and typological framework", *International Journal of Physical Distribution and Logistics Management*, Vol. 48, No. 8, pp. 842–865. https://doi.org/10.1108/IJPDLM-02-2017-0099

Kshetri, N. (2018), "Blockchain's roles in meeting key supply chain management objectives", *International Journal of Information Management*, Vol. 39, pp. 80–89. https://doi.org/10.1016/j.ijinfomgt.2017.12.005

Kumar, S., and Anbanandam, R. (2020), "Impact of risk management culture on supply chain resilience: An empirical study from the Indian manufacturing industry", *Proceedings of the Institution of Mechanical Engineers, Part O: Journal of Risk and Reliability*, Vol. 234, No. 2, pp. 246–259. https://doi.org/10.1177/1748006X19886718

Kuo, Y.-H., and Kusiak, A. (2019), "From data to big data in production research: The past and future trends", *International Journal of Production Research*. 57. 4828–4853. https://doi.org/10.1080/00207543.2018.1443230.

Lämsä, A. M. and Savolainen, T. (2000), "The nature of managerial commitment to strategic change", *Leadership and Organization Development Journal*, Vol. 21 No. 6 pp. 297–306.

Lavastre, O., Gunasekaran, A. and Spalanzi, A. (2014), "Effect of firm characteristics, supplier relationships and techniques used on supply chain risk management (SCRM): an empirical investigation on French industrial firms", *International Journal of Production Research*, Vol. 52 No. 11, pp. 3381–3403.

Lohmer, J., Bugert, N., and Lasch, R. (2020), "Analysis of resilience strategies and ripple effect in blockchain-coordinated supply chains: An agent-based simulation study", *International Journal of Production Economics*, 228 (September 2019). https://doi.org/10.1016/j.ijpe.2020.107882

Min, H. (2019), "Blockchain technology for enhancing supply chain resilience", *Business Horizons*, Vol. 62, pp. 35–45. https://doi.org/10.1016/j.bushor.2018.08.012

Nayernia, H., Bahemia, H., and Papagiannidis, S. (2021), "A systematic review of the implementation of Industry 4.0 from the organizational perspective", *International Journal of Production Research*, https://doi.org/10.1080/00207543.2021.2002964

Neubauer, M. (2018), "Supply chain resilience support in S-BPM", *ACM International Conference Proceeding Series*. https://doi.org/10.1145/3178248.3178263

Oncioiu, I., Bunget, O. C., Türkes, M. C., Capusneanu, S., Topor, D. I., Tamas, A. S., and Rakos, I. S. (2019), "The impact of big data analytics on company performance in supply chain management", *Sustainability Switzerland*, Vol. 11, No. 18, 4864. https://doi.org/10.3390/su11184864

Papadopoulos, T., Gunasekaran, A., Dubey, R., Altay, N., Childe, S. J., and Fosso-Wamba, S. (2017), "The role of Big Data in explaining disaster resilience in supply chains for sustainability", *Journal of Cleaner Production*, Vol. 142, pp. 1108–1118.

Parast, M.M., Sabahi, S. and Kamalahmadi, M. (2019), "The relationship between firm resilience to supply chain disruptions and firm innovation", *International Journal of Logistics Research and Applications*, Vol. 23 No. 3, pp. 254–269.

Patel, B. S., and Sambasivan, M. (2022), "A systematic review of the literature on supply chain agility", *Management Research Review*, Vol. 45, No. 2, pp. 236–260.

Patel, B. S., Samuel, C., and Sharma, S. K. (2018) "Analysing interactions of agile supply chain enablers in the Indian manufacturing context", *International Journal of Services and Operations Management*, Vol. 31, No. 2, pp. 235–259.

Patel, B. S., Tiwari, A. K., Kumar, M., Samuel, C., and Sutar, G. (2020) "Analysis of agile supply chain enablers for an Indian manufacturing organisation", *International Journal of Agile Systems and Management*, Vol. 13, No. 1, pp. 1–27.

Pereira, C. R., Christopher, M., and Lago Da Silva, A. (2014), "Achieving supply chain resilience: The role of procurement", *Supply Chain Management: An International Journal*, Vol. 19, No. 5/6, pp. 626–642.

Pettit, T. J., Croxton, K. L., and Fiksel, J. (2013), "Ensuring supply chain resilience: development and implementation of an assessment tool", *Journal of Business Logistics*, Vol. 34 No. 1, pp. 46–76.

Queiroz, M. M., Pereira, S. C. F., Telles, R., and Machado, M. C. (2021), "Industry 4.0 and digital supply chain capabilities", *Benchmarking: An International Journal*, Vol. 28, No. 5, pp. 1761–1782.

Radhakrishnan, S., Harris, B., and Kamarthi, S. (2018), "Supply chain resiliency: A review. *In* Y. Khojasteh (eds.)", *Supply Chain Risk Management*. Springer, Singapore, pp. 215–235.

Rajesh, R. (2017), "Technological capabilities and SC resilience of firms: A relational analysis using Total Interpretive Structural Modeling (TISM)", *Technological Forecasting and Social Change*, Vol. 118, pp. 161–169. https://doi.org/10.1016/j.engappai.2019.103338

Rajesh, R. (2020), "A grey-layered ANP based decision support model for analyzing strategies of resilience in electronic supply chains", *Engineering Applications of Artificial Intelligence*, Vol. 87, pp. 1–18.

Ralston, P., and Blackhurst, J. (2020), "Industry 4.0 and resilience in the supply chain: A driver of capability enhancement or capability loss?" *International Journal of Production Research*, Vol. 58 No. 15, pp. 5006–5019.

Ramirez-Peña, M., Sánchez Sotano, A. J., Pérez-Fernandez, V., Abad, F. J., and Batista, M. (2020), "Achieving a sustainable shipbuilding supply chain under I4.0 perspective", *Journal of Cleaner Production*, Vol. 244, Article 118789. https://doi.org/10.1016/j.jclepro.2019.118789

Ruiz-Benítez, R., López, C., and Real, J. C. (2018), "The lean and resilient management of the supply chain and its impact on performance", *International Journal of Production Economics*, Vol. 203, pp. 190–202. https://doi.org/10.1016/j.ijpe.2018.06.009.

Sabahi, S. and Parast, M.M. (2020), "Firm innovation and supply chain resilience: a dynamic capability perspective", *International Journal of Logistics: Research and Applications*, Vol. 23, No. 3, pp. 254–269.

Samvedi, A., Jain, V., and Chan, F. T. S. (2013), "Quantifying risks in a supply chain through integration of fuzzy AHP and fuzzy TOPSIS", *International Journal of Production Research*, Vol. 51 No. 8, pp. 2433–2442.

Sangari, M. S., and Dashtpeyma, M. (2019), "An integrated framework of SC resilience enablers: A hybrid ISM-FANP approach", *International Journal of Business Excellence*, Vol. 18, No. 2, pp. 242–268.

Scholten, K., Sharkey Scott, P. and Fynes, B. (2019), "Building routines for non-routine events: supply chain resilience learning mechanisms and their antecedents", *Supply Chain Management*, Vol. 24 No. 3, pp. 430–442.

Shibin, K.T., Gunasekaran, A., Papadopoulos, T., Dubey, R., Singh, M., and Wamba, S.F. (2016), "Enablers and barriers of flexible green supply chain management: a total interpretive structural modeling approach", *Global Journal of Flexible Systems Management*, Vol. 17 No. 2, pp. 171–188.

Simangunsong, E., Hendry, L. C., and Stevenson, M. (2012), "Supply-chain uncertainty: A review and theoretical foundation for future research", *International Journal of Production Research*, Vol. 50 No. 16, pp. 4493–4523.

Singh Patel, B., Samuel, C., and Sutar, G. (2020), "Designing of an agility control system: A case of an Indian manufacturing organization", *Journal of Modelling in Management*, Vol. 15, No. 4, pp. 1591–1612.

Soares, M. C., Ferreira, C. V., and Murari, T. B. (2021), "Supply chain resilience and Industry 4.0: A evaluation of the Brazilian northeast automotive OEM scenario post COVID-19", *Artificial Intelligence Perspective*, Vol. 3, No. 3., pp. 1–12. https://doi.org/10.1186/s42467-021-00010-1

Soni, U., Jain, V., and Kumar, S. (2014), "Measuring SC resilience using deterministic modeling approach", *Computers and Industrial Engineering*, Vol. 74, pp. 11–25. https://doi.org/10.1016/j.cie.2014.04.019

Spieske, A., and Birkel, H. (2021), "Improving supply chain resilience through Industry 4.0: A systematic literature review under the impressions of the COVID-19 pandemic", *Computers and Industrial Engineering*, Vol. 158, article id 107452, pp. 1–22. https://doi.org/10.1016/j.cie.2021.107452

Tukamuhabwa, B., Stevenson, M., Busby, J., and Zorzini, M. (2015), "Supply chain resilience: Definition, review and theoretical foundations for further study", *International Journal of Production Research*, Vol. 53, No. 18, pp. 5592–5623.

Urban, W., Łukaszewicz, K., and Krawczyk-Dembicka, E. 2020. "Application of Industry 4.0 to the product development process in project-type production", *Energies*, Vol. 13, No. 21., pp 1–20. https://doi.org/10.3390/en13215553.

Vilko, J., Ritala, P., and Edelmann, J. (2014), "On uncertainty in supply chain risk management", *The International Journal of Logistics Management*, Vol. 25 No. 1, pp. 3–19.

Wilkesmann, M., and Wilkesmann, U. 2018. "Industry 4.0 – Organizing routines or innovations?" *VINE Journal of Information and Knowledge Management Systems*, Vol. 48, No. 2, pp. 238–254.

Xu, Li Da, Eric L Xu, and Ling Li. (2018), "Industry 4.0: State of the Art and Future Trends." *International Journal of Production Research*, Vol. 56, No. 8, pp. 2941–2962.

Yadav, A. K., and Samuel, C. (2021), "Modeling the barriers of the resilient supply chain: A fuzzy dematel approach", *Journal of Advanced Manufacturing Systems*. https://doi.org/10.1142S0219686722500275

Yadav, A. K., and Samuel, C. (2022), "Modeling resilient factors of the supply chain", *Journal of Modelling in Management*, Vol. 17, No. 2, pp. 456–485.

Yinan, Q., Tang, M., and Zhang, M. (2014), "Mass customization in flat organization: The mediating role of supply chain planning and corporation coordination", *Journal of Applied Research and Technology*, Vol. 12 No. 2, pp. 171–181.

Zhang, Q., Jin, J. L., and Yang. D. (2020), "How to enhance supplier performance in China: Interplay of contracts, relational governance and legal development", *International Journal of Operations and Production Management*, Vol. 40, No. 6, pp. 777–808.

2

Blockchain Using Supply Chain Management for Society 5.0

Parkavi K, Leninisha S, Aishwariya Subakkar, Nivedha M, and J. Vanitha

CONTENTS

2.1 Introduction

Managing a supply chain even for small business isn't easy task. Supply chain management is the art of management. A good flow of service provides correct product at a prefect time with sufficient costs for the customer by which supply chain management is an art of management. Providers, customers, and third-party suppliers are channel followers which makes coordination and association together [1]. Blockchain makes it effortless to route compliance efforts by recording all the required steps which proves it is undoubtedly an incorruptible ledger.

The imprecision in records, fraud and corruption, issues in billing, this intends to analyse the whole chain needs by unbiased third parties, and there is a need to check them simultaneously. Determination of source and building of products, as well as hyper segmentation, are prime drivers behind the change of traditional supply chain. Factors are triggered into huge tech organisations to start a combination of impact to the complete action of the supply chain with high automation approach in a hybrid structure [2]. Since the origination of Bitcoin, Blockchain technology has been the main phenomenon of financial topic. In Blockchain, all the transactions are clearly obvious to all users, and it is traceable in the ledger. It is unchangeable and not reversible at any point during the transactions and does not have any failure. It requires only less time and work force. The automatic updating feature of

DOI: 10.1201/9781003177432-3

the Blockchain ledger one of the advantages. Using the same approach with supply chain will definitely provide great results.

2.2 Related Work

Vecchio et al. [3] proposed a method Agri-BlockIoT with the practical use, which is fully decentralised and traceability with Blockchain solution for Agri-based food supply chain management.

Bouchti et al. [4] proposed an organized mapping analysis of Supply Chain Management using Blockchain approaches. The aim is to examine and identify the state of Blockchain techniques for supply chain management. They have tried to analyse the trends used in SCMs by changing already accessible results.

Anagnostopoulos et al. [5] proposed a particular analysis to check whether it is suitable for Blockchain in the industry of supply chain. They have discussed the crucial Blockchain characteristics with supply chain such as cost and location proof, scalability, privacy, consensus, performance, and information on the Blockchain inspecting the supply chain domain.

Toyoda et al. [6] proposed the Product Ownership Management system based on Blockchain, and they port to explain the available products which intend to duplicate original products of counterfeiters. If the seller does not provide their ownership, even with a genuine product code, consumers can decline the buy of fake products.

Xu et al. [7] discussed the combination of Blockchain in physical and cyber system; as years pass, the data keeps on growing in a large volume. This increases the global trends. They have proposed the major ideas such as in maritime transportation to identify the critical challenge and proposed a detail design and digital identity management policy.

Hegde et al. [8] came up with a solution for the problems faced in agricultural using supply chain using blockchain. The key role is played by customers, traders, and producer cycle. The system consists of several steps. Finally, they concluded that by using Blockchain, numerous problems could be solved, providing care, accelerating efficiency, and reducing wastage.

Fahhama et al. [9] discussed multi-method simulation tool. Supply chain effectively improves the implementation of the organisations. Over the past decade, there has been numerous problems quoted which includes product network design, supplier choice, and product development. They have also discussed the crucial approaches which are automation and connectivity in manufacturing.

Gendy [10] performed an analysis to know about the patient-centred supply chain execution and the approaches adopted. This application works on the domain of Hospital and Pharmacy. The model of SCM is given for integrated hospital management which includes medical records, value processor, customer, quality care, clinical outcome, and society, which will come under subtopics: Care provider, Care delivery system, and Care receiver. These are all combined using technology and information. Enabling and Planning plays an important role which includes Sustainability, Data security, Traceability, Mass customisation, Resource optimisation, Empowered customers, assets management, on demand services, Automation and Smart thinking. Then the author discussed the future research and said that if this study is repeated in forthcoming days, we can use it for other related research works. The author conveys the limitations too. The subjective decisions are used to do final selection of articles. Finally, the author said that the technology enabled supply chain OS will increase efficiency and patient care.

Ma et al. [11] discussed blockchain technology with decentralisation technique to ensure users do not put on faith on the sellers to analyse if products are original. Therefore, manufacturer can use this application to provide real products without controlling the stores, which worked directly. It can decrease the cost and product quality affirmation.

Khalil et al. [12] worked on RFID an anti-counterfeiting approach, and it is used to analyse fake products when a customer buys the product. They have used delicate and inexpensive tag for distribution in wide-range industry has been offered. It does not provide any way for product return scenarios, security verification and applications to test. To know the genuineness and quality of products, they came up with an open architecture product, to maintain the internet-based credit of manufacturing with the help of chemical among various builders. Using this method, the special characteristics of personalised products are represented.

Raj and Sowmiya [13] performed a detailed study in the field of supply chain management using blockchain technology discussing the various types of techniques and methods that are used. They said about the problems that are faced in Blockchain using supply chain management and gave the solutions for the same. They also said about the recent technologies that are used for the supply chain security.

Yue et al. [14] constructed a medical equipment supply chain supervision model based on Blockchain technology, and a medical equipment supply chain supervision system based on Blockchain technology was formed based on the full life cycle supply chain management model.

Sathya et al. [15] discussed the use of Blockchain technology and gave the experimentation in their work with the features of immutability, decentralisation, increased security, and tamper proof for supply chains. Their work involves avoiding falsifying information, external attacks, and corrupting database using food supply chain management systems. They also discussed the issues that are faced in centralised food supply chains and resolved those issues using Blockchain implementation of food supply chain.

2.3 Methodology

In order to manage the supply chain more creditably, a Blockchain-enabled supply chain can help participants to record price, date, location, quality, certification, and other related facts. The availability of this information within Blockchain can increase the traceability of material supply chain, reduce losses from counterfeit and grey market, improve visibility and compliance over outsourced contract manufacturing, and potentially enhance an organisation's position as a leader in responsible with manufacturing.

a. **Understanding Blockchain and its value in today's supply chain:**
   The internet-based technology is used to implement Blockchain. It is popular for its capability to make immutable, encrypted ledgers to validate record and distribute transactions. Blockchain is used to support Bitcoin in the transactions and is also used to function independently from central bank that helps in digital cryptocurrency. So, this technology gives a way so that thousands of computers are linked to networks across the world that helps us to create and distribute the ledger or record of every Bitcoin transaction. This Blockchain technology gives better security when compared with the banking model as we have encrypted transactions and ledgers. Using this technology we remove all the lengthy clearing process and the cost involved to transfer the money for one account to another as via internet we can get instantaneous transmission. Here we have "blocks" of transactions that are validated and "immutable" and are linked together to form a chain in chronological order.

b. **Blockchain's value in today's supply chains**
   Blockchain technology is not used in most of the supply chain in recent times. But the Blockchain technology has inspired and prompted many to initiate pilot projects which includes:
   Walmart tried to verify the transactions and tried to maintain perfection and efficiency of record by testing a system that traces pork in China and analysed in the US.
   Maersk and IBM use Blockchain technology to improve efficiency of the process by working on cross-border and cross-party transactions.
   BHP substitute spreadsheets for tracing samples internally and externally from a scope of providers by analysing a Blockchain solution.
   A UK start-up, Provenance adapted Blockchain technology to trace food by raising $800,000. It piloted by tracing tuna in the supply chain of Southeast Asia.
   "Can Blockchain technology add value to supply chains?" is the question raised by the authors who are not sure of the applications related to supply chain.

In reality, nowadays supply chains have good data that can be transferred at the speed of real time. Blockchain technology adds value to three areas:

- It replaces slow and manual processes – Even the current supply chain can handle large and complex datasets and many of the processes; they are slow and depend on paper in lower supply tiers. It is most common in shipping industry.

- Increasing traceability – The change made is high regulatory and consumer demand for provenance of information. High-cost quality issues such as reputational damage, recalls or the loss created due to black or grey market products can be mitigated by improving traceability. Value-creation opportunities can be offered by simplifying complex supply base offers.

- Reducing cost of supply chain IT transaction – The benefit is more theoretical. Bitcoin validates each block by paying people, and a fee need to be included in the proposal if people propose a new block. This cost can be prohibitive as the scale may be staggering. There will be a significant rise in demand for an important component of Blockchain's distributed-ledger approach, which is data storage. Moreover, creating and maintaining more copies of datasets, especially in permission-less Blockchains, will not be practical in supply chain environment.

c. **Key characteristics of Blockchain:**

The Blockchain can streamline administrative processes and reduce costs by enabling an in effect audit of supply chain data.

a. **Distributed Ledger**

Digital Blockchain is known as distributed ledger. They utilise individualistic computers to share, synchronise transactions, and record in their corresponding ledgers instead of storing the data values in conventional manner. Blockchain manages data into blocks, which are then combined jointly in a mode called append. They are the building blocks that enhance recording of link and transfer "value" peer-to-peer, where the need for a centrally coordinating entity doesn't occur. "Value" refers to any history of asset values or personal data. DLT can fundamentally change Society 5.0 by making it more reliable and efficient

b. **Cryptography**

Blockchain ledger related to cryptography which acts as key of security. Using encrypted data, each transaction is recorded on the Blockchain. Any user can access their own information and securely buy and sell crypto, using public and private keys provided to them. A Blockchain is an increasing list of records, called blocks, which are connected together using cryptography. Each block has a cryptographic hash of the former block, transaction data value, followed by a timestamp.

c. **Consensus**

Consensus is an algorithm where decentralised record keeping more like centralised database. It's an automatic action to confirm there only exists one single valid copy of record shared by all the nodes. In order to be valid transaction, all parties have to agree. Without consensus, no new blocks are developed or changes are made. This means all parties agree to the change and know when a change is being made. All of them in the chain agree that all transactions are valid. In the case of supply chain management for Society 5.0, Blockchain technology can be used to bring consensus to numerous transactions which includes payment, warehouse management, delivery and transportation.

d. **Smart contracts**

A Blockchain technology with smart contract termed as software program used to carry out a digital agreement powered by Blockchain. In order to block fraudulent scheme, smart contracts with Blockchain can be combined with the supply chain. Basic supply events, such as payment, serial number mismatch, are automatically triggered using programming. Smart contracts with Blockchain technology help to express, verify, and carry out the terms as per the contract agreed by the parties. Blockchain qualifies the agreements and transactions elaborated in the contract to be taken over without depending on a legal system or external enforcement.

e. **Why Blockchain is introduced to supply chain operations?**

i. **Traceability**

Blockchain-based traceability has the ability to recognise fraudulent transactions, tracking product origin, and supply chain schemes at the same time, which can reduce traditional paperwork process. It has three key advantages: ability to support condition control systems, decreases risk, and enlarge supply chain visibility. By keeping a record of the entire production and distribution history, suppliers are able to act immediately to any controversy. Traceability has been aggressively forwarded in the supply chain management. It is highly utilised for decreasing past issues, reducing damages, and enhancing management challenges as well as checking quality management. It is, however, crucial to ensure the data from manufacturing through disposal of all components numbering numerous tens of thousands and to check laws and regulations that vary with the times. On the other hand, globalisation is advancing, while delivery time and cost competition are emphasised in recent years, so the significant of traceability keeps on growing (Figure 2.1).

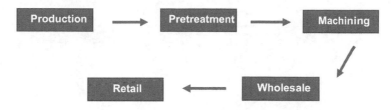

FIGURE 2.1
Traceability in supply chain management.

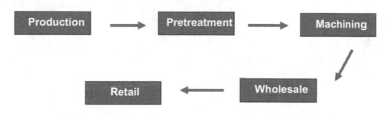

FIGURE 2.2
Transparency in supply chain management.

ii. **Transparency**

Blockchain has a key characteristic which is transparency. It helps us to join the network and view and follow the results of the network. The transparency of Blockchain gives users a possibility to look through the past transactions in case of crypto currency. Blockchain can qualify numerous transparent and accurate tracking in the supply chain. Available to trace assets from production to delivery or use by end user, organisations can digitise physical assets and create a decentralised fixed record of each contract. Supply chain transparency needs companies to know what upstream undergoing in the supply chain and to pass on this knowledge both internally and externally. More consumers are demanding it because of one reason where process has become increasingly important. Supply chain transparency enable suppliers instantly acquire and process the data they need and pass these time savings on to their clients. By selecting to be an accessible partner, from the contesting to qualify existing business and probably even gain new clients suppliers can stand out uniquely (Figure 2.2).

iii. **Supply chain finances**

Solutions based on technology which aims for low financing price and enhance business strategies for buyers and sellers together in sales transaction is termed as supply chain finance. In recent

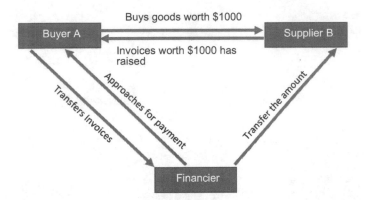

FIGURE 2.3
Supply chain finances in supply chain management.

times, there has been a great deal of interest in Blockchain and supply chain finance solutions. These two combined together can multiply the efficiency of invoice processing and contribute numerous transparent and safe transactions while increasing the efficiency of invoice processing. Supply chain finance employs the best when the buyer has good credit rating than the seller and can accordingly source capital from other financial providers or banks at a lower cost. This benefit lets buyers arrange better terms from the seller, which helps in extended payment schedules. To obtain fast payment from the intermediary financing body, the seller can unlade its products more quickly. Supply chain finance, often mentioned as "reverse factoring", or "supplier finance", strengthens collaboration between buyers and sellers (Figure 2.3).

2.4 Conclusion

Blockchain helps us to reduce the issues faced by supply chain. It greatly helps in the payments, transparency among the participants in the supply chain for Society 5.0. The supply chain is a complex system which comprises many components and stakeholders. Traceability, transparency, and supply chain finance throughout the supply chain are attainable through this chapter. Also, key features of Blockchain can help eliminate costly delays and storage space. It can enhance transaction and automatise systems. Challenges and benefits in the field of supply chain for Society 5.0 using Blockchain helps us to understand the content more in an effective way. Many have taken steps toward the implementation of Blockchain-enabled supply chain, with some accurate results.

References

[1] H. Stadtler. (2015). Supply chain management: An overview. In *Supply Chain Management and Advanced Planning* (pp. 3–28). Berlin, Heidelberg: Springer.

[2] G. Kelechi, C. M. Akujuobi, M. N. Sadiku, M. Chouikha, S. Alam. (2020). Internet of things and Blockchain integration: Use cases and implementation challenges. In *Business Information Systems Workshops: BIS 2019 International Workshops*, Seville, Spain: Springer Nature, June 26–28, 2019. Revised Papers, Vol. 373, p. 287.

[3] M. Vecchio, M. Pincheira Caro, R. Giaffreda, M. Salek Ali. (2018). A practical implementation: Blockchain-based traceability in agriculture-food supply chain management. In *Proceedings of the International Conference on IoT Vertical and Topical Summit on Agriculture Tuscany*, Italy, 8–9 May.

[4] A. El Bouchti, H. Bouayad, Y. Tribis. (2018). A systematic mapping study on supply chain management based on blockchain. *International Workshop on Transportation and Supply Chain.*

[5] D. Anagnostopoulos, T. Varvarigou, A. Litke. (2019, January). *Architectural Elements and Challenges Towards a Global Scale Deployment: Blockchains for Supply Chain Management.* Multidisciplinary Digital Publishing Institute (MDPI).

[6] K. Toyoda, P. Takis Mathiopoulos, I. Sasase, T. Ohtsuki. (2017, June). *A Novel Blockchain-Based Product Ownership Management System (POMS) for Anti-Counterfeits in the Post Supply Chain.*

[7] L. Xu, L. Chen, Z. Gao, Y. Chang, E. Iakovou, W. Shi. (2018). *Binding the Physical and Cyber Worlds: A Blockchain Approach for Cargo Supply Chain Security Enhancement.* IEEE. 978-1-5386-3443-1/18.

[8] B. Hegde, B. Ravishankar, M. Appaiah. (2020). Agricultural supply chain management using Blockchain technology. In *Proceedings of the International Conference on Main streaming Block Chain Implementation (ICOMBI)*. IEEE. 978-93-5406-901-7.

[9] L. Fahhama, A. Zamma, K. Mansouri, Z. Elmajid. (2017). *Towards a Mixed Met Hod Model and Simulation of the Automotive Supply Chain Network Connectivity.* IEEE. 978-1-5386–0875-3/17/$31.00.

[10] A. Wassimal Gendy. (2019). *Review on Healthcare Supply Chain.* IEEE. 978-1-7281-5052-9/19/$31.00.

[11] J. Ma, S.-Y. Lin, X. Chen, H.-M. Sun, Y.-C. Chen, H. Wang. (2020, February). *A Block Chain Based Application for Product Anti-Counterfeiting.*

[12] G. Khalil, R. Doss, M. Chowdhury. (2020, March). *A New Novel RFID Based Anti-Counterfeiting Scheme for Retail Environment.*

[13] Y. Raj, B. Sowmiya. (2021, February 26). *Study on Supply Chain Management Using Blockchain Technology.*

[14] Y. Yue, X. Fu. (2020, December 17). *Research on Medical Equipment Supply Chain Management Method Based on Blockchain Technology.*

[15] D. Sathya, S. Nithyaroopa, D. Jagedeesan, I. Jeena Jacob. (2021, March 31). *Blockchain Technology for Food Supply Chains.*

Section 2

Barriers in Adoption

3

Evaluating the Barriers of Adoption of Society 5.0 by Emerging Economies

Neelesh Thallam, Snigdha Malhotra, and Deepali Ratra

CONTENTS

3.1 Introduction

An individual's social system may be characterised as one in which a group of people are continually in touch with one another while operating within the confines of political and cultural boundaries. Throughout history, numerous civilisations have emerged and have been classified according to their access to resources, power, and prestige.

Human growth has progressed through several periods of society. Society 1.0 was dubbed the "non-productive economy" because its inhabitants were nomads. They hunted and gathered. As people began agriculture, cultivation, and related agricultural activities, Society 2.0 was dubbed "Farming Society". Society 3.0 is referred to be an "industrial society" due to the establishment of factories and technological advancements. This society altered the way people thought and introduced the concept of social class systems. Additionally, throughout this time period, people began to earn rights. Then there is Society 4.0, sometimes identified as "information society", which has revolutionised access to information and interaction between individuals through the use of ICTs. Society 5.0 is a stage in this evolution that strives to create a thriving human-centred society (Narvaez Rojas et al., 2021). After passing through hunting society, the agriculture focused society, and then transitioning through industrial era, and coping with an information age, we are now entering an era of Internet of Things (IoT) & Big data society called Society 5.0 (Shiroishi et al., 2018).

Japan's notion of a technology-driven, human-centred society arising from the Fourth Industrial Revolution is called Society 5.0 (Yuko, 2017). This notion began in 2016. Previously, data was saved in the cloud and then retrieved and evaluated by employees, but in society 5.0, data will be kept in cyberspace and then analysed by artificial intelligence (AI). This studied data will be reintroduced into the real world. Data, information, and knowledge are important components of this integration's success. In Society 5.0, humans will have access to AI-derived knowledge; this knowledge will be able to stimulate innovations in primary, secondary, and tertiary sectors as well as completely restructure the industry structure. It is the goal of Society 5.0 to create a more human-centred society that will accelerate economic development while also addressing social challenges (Phuyal et al., 2020). The objective is to provide a better standard of living and to address current societal concerns through the integration of virtual and material world (physical space) as well as to have a digitally prepared workforce (Ablyazov et al., 2018).

According to Arsovski (2019), to progress towards the achievement of Society 5.0, Japan must strengthen its research and development, ensure the reinforcement of systems that support the realisation of a super-smart society, and enhance and practically apply basic technologies that support a super-smart society. He also emphasised the importance of cultivating and securing the following human resources that can contribute to a super-smart society. (i) Engineers have expertise in AI who are conversant with cutting-edge technologies (Moore et al., 2017); (ii) data scientists (iii); cybersecurity experts; (iv) Individuals with an entrepreneurial spirit; and (v) cultivation of the characteristics and abilities necessary for survival in an age of superintelligence.

Society 5.0 is a strategy adopted by Japan for its economic growth is not limited to Japan (Onday, 2019; Shiroishi et al., 2019). According to a study by Zengin et al. (2021), the solutions generated by Society 5.0 can be used to address the world's challenges and achieve the 17 sustainable development goals (SDGs), which broadly include smart technology for agriculture and food, early cautioning systems, virtual knowledge systems, women employment, smart grid systems, I-constructions, global innovations and ecosystems, smart cities, and the use of metrological and other advanced technologies (Fukuyama, 2018). For a long period of time, SDGs have been an integral part of Indian tradition. India's commitment to the SDGs can be gauged by the current government motto "Sabka Saath Sabka Vikas", which translates as "Collaborative efforts and Inclusive development". These objectives will be impossible to attain without a high degree of governance at all levels. The Indian government is collaborating with civil society, the private sector, and other stakeholders on this initiative. In India, NITI Aayog plays a critical role in achieving these aims (Nair et al., 2021). With the Digital India Mission and other e-governance initiatives, India has enormous potential for Society 5.0. The Minister of Economy, Trade, and Industry of Japan urged Indian software sector to participate in developing solutions for Society 5.0 in 2018. Though Society 5.0 can be a boon for India, it can address the following issues: it can address agricultural problems through inventions and automations; it can ensure the availability of affordable services to senior citizens; it can ensure good governance and maximum governance; it can assist in eradicating illiteracy through digitisation and cloud computing; and it can address traffic, medical health care, and manufacturing issues through robotics, big data, and artificial knowledge engineering.

However, it has the potential to improve awareness of concerns such as rising inequality, increased demand for energy and food, irresponsible internet and social media use, and environmental degradation. To put it more succinctly, it is an integrated society that balances economic advancement with social problem-solving. New technologies like the IoT, robotics, AI, and big data are being developed, and Society 5.0 is a new society that incorporates these technologies.

Efforts to minimise greenhouse gas (GHG) emissions, enhance production and reduce food waste, balance the expenses of an ageing population, and promote sustainable industrialisation, income redistribution, and regional inequality correction must be made by emerging countries. Achieving economic success while addressing social problems has proved challenging, but Industry 4.0 and associated technologies have the potential to alleviate many of those concerns.

The goal of Society 5.0 is to establish a society in which individuals are free to pursue their passions and live their lives to their utmost potential. As a result, economic progress and technological advancement exist, not just to enrich a few individuals, but to benefit all economies as a result of their existence. Furthermore, although Society 5.0 is located in Japan, its mission is not confined to the advancement of the country's economy. Without a doubt, the ideas and technologies developed here will be useful in solving worldwide social concerns in the future.

Our research adds the following value to the problem. First and foremost, we investigate Society 5.0 using a multidisciplinary approach, focusing on societal impact, technological change, and implementation challenges for emerging economies. Further, this chapter attempted to analyse social media trend towards Society 5.0 by scrapping tweets and social media data. Also, the study uncovers the challenges emerging economies like India would be confronting while implementing Society 5.0 and bringing the revolution in their economies.

The majority of research employ a variety of discourses to assess the issues posed by the Fourth Industrial Revolution and to investigate ways to comprehend the disruptions generated by it. Despite the growing importance of automation in the future, no emphasis has been placed on adoption techniques. The current study was motivated by this stringent requirement in emerging countries to identify the critical issues associated with implementing Society 5.0 in the era of Industry 4.0. As a result, the following research questions arise:

- Is it possible for rising economies to determine what is preventing Society 5.0 from being implemented in the Fourth Industrial Revolution era?
- How can we strengthen governments, corporations, and employees to achieve Society 5.0 in the context of Industry 4.0?
- How might these difficulties be classified and organised with the assistance of industry and academia?

In this backdrop, the current study applied a text mining methodology to examine and comprehend the obstacles faced by growing economies, like India, in implementing Industry 4.0 and adopting Society 5.0. As literature on the issues confronting Society 5.0 is scarce, we gathered semantic statistics from Twitter and other social media platforms. A word cloud was created to illustrate the nature of discussion around Society 5.0 on social media platforms. Later, opinion mining was used to analyse the difficulties based on interviews with industry and academic experts. The challenges were gathered through semi-structured interviews. Classification and organisation of these difficulties were accomplished through the use of opinion mining. These issues were discussed in greater detail from the perspective of a rising economy.

The remaining portions of the investigation are as follows: Section 3.2 details the research methods used to conduct the study, which includes Twitter scraping, text mining, sentiment analysis, and opinion mining. Section 3.3 explains the findings. Sections 3.4 and 3.5 discuss the study's shortcomings and conclusions, respectively.

3.2 Research Methodology

The technique of the study is separated into two stages. Twitter scraping was performed in the initial step, followed by text mining and sentiment analysis.

We identified specific emotions and trending topics in society 5.0 because of the investigation. To gain a improved comprehension Society 5.0's difficulties, semi-structured interviews with ten industry and academic specialists were undertaken. The specialists were told on the research study's objectives. Once they consented to participate in the study, a preferred platform for conducting the interview (zoom/Microsoft Teams) was chosen. The interviews were taped, and the issues raised by the interviewees were examined subsequently by the study team.

3.2.1 Scraping Twitter

Twitter is a popular live microblogging service that allows users to post 140-character messages. Users post tweets on a range of issues that are relevant to their everyday life. Twitter is a fantastic instrument for eliciting widespread public opinion on a wide variety of issues and circumstances. Opinion mining and natural language processing-related tweets make up the bulk of the core corpus for sentiment analysis.

To put it another way, Twitter's data dissemination strategy is simple and efficient, according to Antonakaki et al. (2021). However, unlike other OSNs, it is primarily utilised for the dissemination of news, which is a distinct feature that sets it apart (Kwak et al., 2010). As a result of the existence of Twitter accounts for a variety of different types of organisations and individuals, the social media platform boasts a diversity of different Twitter users. With the assistance of these entities, as well as individual user accounts, Twitter has developed into an extremely attractive research topic in a range of domains.

3.2.2 Mining of Text

Text mining is the process of discovering previously unknown material using computers that automatically extract information from a range of unstructured text sources (Tan, 1999). Text mining data sources are generally unstructured since they use human grammar, or what is commonly referred to as natural language.

Pre-processing data is the initial stage of text mining that converts unstructured input into structured data. The following are the stages involved in data pre-processing: slang word conversion (converting non-standard words to standard words), filtering (removing unimportant words using stopwords), tokenisation (breaking words into more meaningful or token words) and stemming (converting affixed words to root words) (Tan et al., 2016).

A word cloud is a visual representation of text that is frequently used in text documents. The purpose of a word cloud is to produce a two-dimensional representation of a text by charting frequently occurring phrases. The most commonly occurring words in a manuscript are represented by a word cloud. The frequency with which common words appear demonstrates the word's frequent appearance in documents.

As a collection of Web-based applications built on Web2.0's conceptual and technical foundations, social media makes it easier for anyone to create and share their own content. Even while companies utilise social networking sites to find and engage their consumers, research shows that social networking is associated with a drop in productivity.

Due to the ease with which social media may be shared with the public, private information may enter the social arena, which can be negative. The study made use of a variety of social media platforms, including Twitter and Facebook. We conducted sentiment analysis using the content uploaded on the websites. "Society 5.0" and "Society 5.0 emerging economy" were used as search terms.

3.2.3 Analysis of Twitter Sentiment

Emotional comments and tweets can be identified and exploited to provide valuable indicators for a variety of different reasons (Sarlan et al., 2014). Positive and negative words may be used to describe a person's mood, according to some researchers. Natural language processing methods may be used in sentiment analysis to quantify a stated opinion or sentiment in a sample of tweets (Feldman, 2013).

There are many other ways of looking at it, but it's often used to describe the process of removing polarity and subjectivity from semantic orientation. It is also referred to as sentiment analysis (Diyasa et al., 2021). When it comes to automatically extracting sentiment, there are two primary approaches to this problem: lexicon-based extraction and machine-learning-based extraction.

Lexicon-based strategies are ways that make use of predefined list of words or an alphabetical listing of terms, each of which is connected with a certain attitude, to achieve their objectives. Regardless of whether or not the lexical approaches were developed within the setting in which they were developed, they all have as their goal the determination of the orientation of a document in accordance with how sentences or phrases included within the document were structured semantically (s). Additionally, it is stated that the objective of a lexicon sentiment is to identify words in a corpus that transmit opinion and then forecast the opinion expressed in a document. The following lexical methods are based on a fundamental paradigm:

1. Prior to posting each tweet, remove any punctuation.
2. Set the overall polarity score (s) to zero (s=0).
3. Determine whether a token is contained in a dictionary and, if it is, continue. If the token is positive, the value of s will be positive (+); otherwise, the value of s will be negative (–). (–)
4. Consider the post's overall polarity.

If s is greater than the threshold, the tweet will be considered positive; if s is lesser than the threshold, the tweet will be considered negative.

The learning-based approach seems to have a number of benefits, one of which is its capacity to modify and refine trained dummies in order to meet specific needs and circumstances, which is one of the most noteworthy advantages. The paucity of labelled data, as a consequence of the restricted application of the fresh data approach that was used to label data, may make labelling data prohibitively costly or even impossible for some occupations, although labelling data is not prohibitively expensive or even impossible for other activities.

3.2.4 Sentiment Analysis Techniques

Using semantic ideas of entities derived from tweets, it is feasible to examine the connection between the general emotion polarity of a group of things by analysing the semantic concepts of entities (Liu, 2012). Whether a phrase or statement is positive or negative may be described in its most basic form, that is, whether it is positive or negative. Non-verbal learning (NLP) approaches are based on the principles of machine learning, more specifically statistical learning. A generic learning algorithm is used in combination with a large sample of data, referred to as a corpus, in order to learn the rules of a language. In computer science, human-computer interaction (HCI) is a branch that is concerned with training computers to draw meaning from human language and input in order to interact with the actual world. NLP is an abbreviation for natural language processing. CBR, on the other hand, is a way for doing sentiment analysis. It is not the only approach. CBR is differentiated by its ability to remember previously addressed difficulties and apply those answers to a fresh set of issues that are very similar to the ones that were previously handled.

3.2.5 Programming Interface for Applications

Alchemy's Application Programming Interface (API) beats their rivals in terms of quality and volume of things that can be retrieved from the database. The Python Twitter API is gradually built up from the accumulated tweets. Python can automatically determine the frequency with which messages are retweeted once every 100 seconds, rank the top 200 messages according to their retweet frequency, and save the results to a specified database.

3.2.6 Mining Exploration

These naturally occurring terminologies include natural language processing, text mining, and computational linguistics, all of which are used to describe the wide domains of natural language processing and text mining. Opinion mining is the automated examination of textual expressions of attitudes, beliefs, and emotions. Although the term "sentiment" refers to a perspective or attitude that is motivated by emotion rather than logic, it is not always applied in this sense. A wide number of industries may benefit from opinion mining, according to the author of the article. This includes accounting and law; research; entertainment; education; technology; politics; and more. It was common for people to express their views and opinions on social media platforms in the early days of the internet.

During a semi-structured interview, experts were asked to share their thoughts with the group. For developing nations, progressing toward Society 5.0 while still embracing the Fourth Industrial Revolution is fraught with challenges. This data assisted the writers in identifying difficulties in the execution of Industry 4.0 efforts. The interview session included experts from both the private and academic sectors. A list of codes was created and organised according to the identified difficulties.

3.2.7 Participants

Fifteen senior analysts, bureaucrats, supply chain professionals, economists, and academicians were contacted for the interview, and ten consented to participate. LinkedIn was used to contact the responders. Shortlisted profiles were discussed among the authors. A note summarising the study's aims was sent to their email address. The interviews were scheduled in advance, and permission was obtained to record the interview. A reference guide was created for the interview that was referred to throughout the conversation. Because the purpose of the interview was to speak with more experienced professionals, the vast majority of those who took the survey had been working in their respective fields for more than 10 years (Table 3.1).

3.2.8 Interviewing Procedures

We designed an interview guide with open-ended questions and utilised clarifying inquiries as needed during the interviews. According to the contributing experts, the questions were aimed to analyse the problems associated with getting toward Society 5.0 acceptance.

- What significance does Society 5.0 have for rising economies such as India?
- What are the prerequisites for implementing Society 5.0 in any economy?
- Countries are establishing smart cities; are we on the verge of achieving Society 5.0?
- Do you believe emerging economies have fully embraced the Industry 4.0 transformation and are fully prepared to implement Society 5.0?

TABLE 3.1

Participant Summary

Participants	Organisation	Designation
1	Schneider	Supply chain head
2	Smart cities	Advisor
3	Gartner	Vice HR, analytics
4	EY	Senior HR analyst
5	Leela hostels	Senior analyst
6	University	Professor
7	University	Dean
8	HCL	IT head
9	IBM	Senior business analyst
10	Avesta Computers Pvt Ltd	Senior business analyst

- Society 5.0 intends to address a number of issues by moving well beyond economic digitalisation. What are your thoughts on this in relation to India?
- Society 5.0 prioritises human needs and embraces automation, even as we witness employment losses due to automation. Isn't Society 5.0 an existential threat to humanity?
- Mention the importance of progressive adoption throughout countries, as well as the importance of providing social security coverage to threatened jobbers, and so on. How can countries prepare for the future?

The interviews lasted for 30–45 minutes each. The responses were recorded with permission from participants. Figure 3.1 depicts the many stages of the research process.

3.3 Discussion

This chapter examines the problems of Society 5.0 in emerging economies on a sector-by-sector basis. We performed comprehensive research and in-person interviews with industry experts to identify sector-specific characteristics that will contribute to the successful implementation of Society 5.0 in India. Additionally, we scraped data from Twitter using an API and ran sentiment analysis on the data, as illustrated by the accompanying visuals (Figure 3.2).

An individual tweet's sentiment is represented in Figure 3.2 by a circle. The location of the circle indicates how the tweet's content elicits an emotional

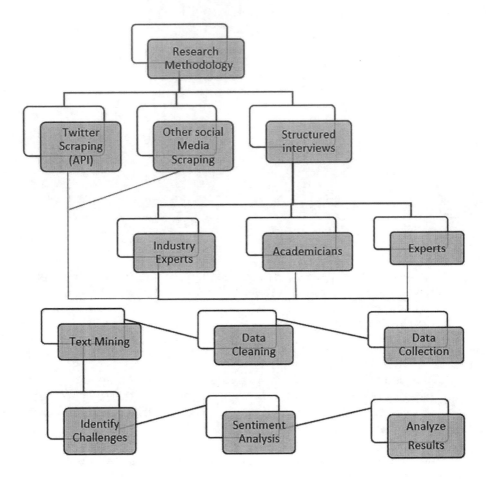

FIGURE 3.1
Research process.

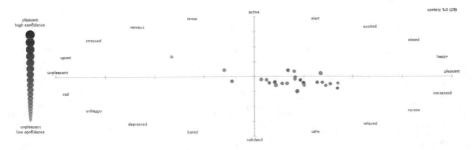

FIGURE 3.2
Sentiment visualisation.

response. Figuratively, negative tweets are represented by varying shades of gray circles (see Figure 3.2). Sedate tweets shown at the bottom of the graph by darker circles, whilst active tweets are displayed at the top by brighter circles. Individuals in developing countries appear to be peaceful, secure, and tranquil about Society 5.0, as seen by a huge quantity of pleasant and sedate tweets (Figure 3.3).

Pleasure and arousal are used as dividing elements in order to divide emotion into an 88-grid. The number of tweets that fall inside each grid cell is calculated to determine determining the colour of resulting grid cell: more tweets than average and fewer than average tweets. There are no tweets in the white cells. In other words, there is a general acceptance of Society 5.0 and its implications (Figure 3.4).

The emotional zones of Upset, Happy, Relaxed, and Unhappy are shown with frequently used phrases. Words that are frequently used have larger font sizes. As demonstrated, no disparaging language was used, and the most often used terms included "social", "happy", and "hope" (Figure 3.5).

To highlight the number of tweets sent over a period of time, a bar chart is used to depict the tweets sent during that period. At the top of the figure, positive tweets are emphasised in green, whereas negative tweets are

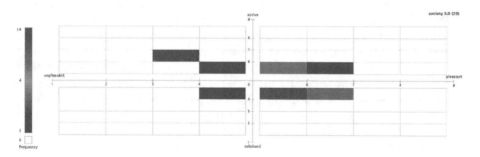

FIGURE 3.3
Heatmap.

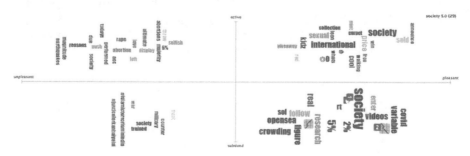

FIGURE 3.4
Tag cloud.

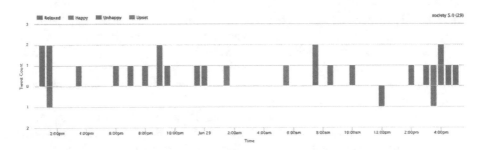

FIGURE 3.5
Timeline.

highlighted in blue towards the bottom. This suggests that majority of tweets were sent during the day's early morning hours.

We were able to discover the consequences of implementing Society 5.0 in emerging countries, like India, by breaking them down by sector using the aforementioned study. The survey's findings are summarised in Table 3.2. Describe how Society 5.0 is being deployed in developing economies on a sector-by-sector basis.

Additionally, the interviews enabled us to highlight the broad obstacles economies face in implementing Society 5.0. Experts emphasise the important difficulties that these economies are adopting Industry 4.0 methods and moving toward automation. Organisations and governments must be prepared to adopt Society 5.0 in terms of infrastructure, cost, and skilled workers. Table 3.3 contains a list of difficulties obtained from expert engagement.

The study indicates that the public and private sectors will need to collaborate to create the optimal climate for growth in emerging economies. A clear emphasis has to be laid on labour market structures, competition, talent and skills development, IT infrastructure, sufficient funding, data safety, security and policy regulations. An increasing skill gap is a major concern for emerging economies and strategies for upskilling current talent will need specific attention. Government support for successful implementation would be evident through supportive regulatory policies and technological support. Economies would be required to emphasise on building better IT infrastructure and invest in data security policies. These approaches may be able to assist Society 5.0 in maximising the benefits of data-driven innovation while also revitalising productivity and growth.

3.4 Relevance for India

With the Digital India Mission and other e-governance initiatives, India has enormous potential for Society 5.0. NITI Aoyog has released a discussion

TABLE 3.2

Sector-Wise Implementation of Society 5.0 in Emerging Economies

Sector-Factor	Description
Medical and healthcare	As the population of adults over the age of 65 continues to grow at an alarming rate, health care and social security expenditures are rising, as well as the need for elderly care. Society 5.0 will tackle the problem of health difficulties by linking and exchanging information across medical data users, and robots will take care of the old population, among other things.
Infrastructure	Roads, bridges, tunnels, and dams will be inspected and maintained with the help of artificial intelligence and robotics.
Mobility	It will promote use of autonomous transportation system and it will also improve distribution and logistics efficiency by introducing innovative technologies.
Finance/fintech	Improving efficiency of banks and promoting cashless transactions.
Education	Ensuring access, affordability, quality, and uniformity. Illiteracy can be eradicated with the help of digitisation and cloud computing.
Food and agriculture	Agriculture problems can be solved with mechanisation and automation.
Energy and resources	Waste to energy plants and efficient management of resources using AI.
Service sector	Increase productivity.
Economic advancement and social problems	Make use of a wide range of new technology in order to accomplish both economic growth based on the United Nations' Sustainable development goals as well as answers to some of the most pressing social issues that now plague our world.
Environmental and climate changes	Decreasing greenhouse gases and sustainable usage of resources.
sustainable industrialisation	Greener and efficient means of production, elimination of difficult labour.
Smart cities	Better connectivity, mobility, and security.
Government	Proactive and predictive service delivery, elimination if middlemen, feedback mechanisms.

paper titled "AI for All", which examines the application of AI in a variety of sectors, including education, health, and agriculture.

Society 5.0, a new society that incorporates new technologies such as the Internet of Things, robotics, AI, and big data into all industries and social activities, is envisioned as a human-centred society that strikes a balance between economic advancement and the resolution of social problems.

To minimise GHG emissions, increase food production and reduce food waste, alleviate the expenses of an ageing population, promote sustainable industrialisation and regional inequality rectification, India must make significant efforts in all these areas. However, within the current social system, achieving economic improvement while also addressing social issues has proven difficult. On the other hand, Industry 4.0 and related technologies may be utilised to address a number of these issues, among others.

TABLE 3.3

Comprehensive List of Challenges

Social	Community uncertainty
	Policy and regulation problems
	Infrastructure and energy consumption
Economic	Cost and expenditures
	City architecture
Technical	Inadequate smart technology interactions and misconnections
	The present approaches are being redesigned in order to provide more fresh findings
	Insufficient AI applications
	Problems of data management and security
Government	Innovation policy by government
	Insufficient funds and grants
	Lack of IT infrastructure
	Lack of entrepreneurial policies
Organisational	Creating change enabling work culture
	Link between various departments
	Honing existing skills and fostering new ones
	Widening skill gap
	Employee resistance
	Digital capabilities and digital orientation
	Digital innovation culture

3.4.1 India's Challenges

- Growing inequality
- Increasing energy and food consumption
- Improper use of the internet and social media
- Environmental Pollution Society 5.0 has the potential to be a benefit for India, as it can handle a variety of issues.
- Mechanisation and automation can assist tackle agriculture's problems.
- Through the use of technology and innovation, senior citizens can receive services at cheap prices.
- Blockchain technology has the potential to achieve both good and maximal governance.
- With the use of digitisation and cloud computing, illiteracy can be abolished.
- Big data, robotics, and AI have the potential to tackle problems related to traffic, medical health care, and industrial and service sector productivity.

3.5 Research Limitations and Future Directions

The research used only social media posts and Twitter data, which were then verified through interviews with specialists from a range of businesses who held senior management positions in their respective disciplines. The perspectives and experiences of workers at all levels and from all corners of the globe may be worth examining in the future, opening a vast field of study and research possibility. Additional study could be conducted utilising multi-criteria decision-making methodologies in order to amass a more comprehensive set of facts. Multi-criteria decision-making (MCDM) techniques may be used to prioritise the issues confronting policymakers. Despite some of the study's limitations, the findings provide valuable information that should be of interest to academics, academicians, and professionals worldwide, particularly in emerging countries like India.

3.6 Conclusion

The globe is currently undergoing a tsunami of change and is accelerating its adoption of technology advancements. These developments brought about by the Fourth Industrial Revolution are both useful and detrimental to humanity. In an age of AI-enabled systems, society's evolution will be entirely dependent on humans' desire to use this technology. Organisations' old tactics are compelled to deviate and pursue a new path. When humans confront and accept new technology transformations, organisations are compelled to do the same. However, as our reliance on technology increases, various issues such as privacy concerns, a lack of security, and infrastructure concerns arise. The report examines emerging economies' issues (government, human resources, technical, economic, and social). Organisations and policymakers can use the challenges to develop strategies for a more effective implementation of Society 5.0. Finally, Industry 4.0 is a must for the next generation and will alter the lifestyles of Society 5.0.

References

Ablyazov, T., Asaturova, J., & Koscheyev, V. (2018). Digital technologies: New forms and tools of business activity. In *SIIS Web of Conferences* (Vol. 44, p. 00004). EDP Sciences.

Antonakaki, D., Fragopoulou, P., & Ioannidis, S. (2021). A survey of Twitter research: Data model, graph structure, sentiment analysis and attacks. *Expert Systems with Applications*, 164, 114006.

Arsovski, S. (2019, December). Quality of life and society 5.0. In *13th International Quality Conference* (397–404), ISSN 2620-2832.

Diyasa, I. G. S. M., Mandenni, N. M. I. M., Fachrurrozi, M. I., Pradika, S. I., Manab, K. R. N., & Sasmita, N. R. (2021, May). Twitter sentiment analysis as an evaluation and service base on python textblob. In *IOP Conference Series: Materials Science and Engineering* (Vol. 1125, No. 1, p. 012034). IOP Publishing.

Feldman, R. (2013). Techniques and applications for sentiment analysis. *Communications of the ACM*, 56(4), 82–89.

Fukuyama, M. (2018, August). Society 5.0: Aiming for a new human-centered society. *Japan SPOTLIGHT*, pp. 47–50. https://www.jef.or.jp/journal/pdf/220th_Special_Article_02.pdf

Kwak, H., Lee, C., Park, H., & Moon, S. (2010, April). What is Twitter, a social network or a news media? In *Proceedings of the 19th International Conference on World Wide Web* (pp. 591–600). IEEE.

Liu, B. (2012). Sentiment analysis and opinion mining. *Synthesis Lectures on Human Language Technologies*, 5(1), 1–167.

Moore, R., Narsalay, R., Seedat, Y., Sen, A., & Chen, J. Y. J. (2017). *Combine and conquer: Unlocking the power of digital industry X. 0.* https://repository.up.ac.za/bitstream/handle/2263/82665/Chen_Combine_2017.pdf?sequence=1 [Accessed on 13 Jan 2022].

Nair, M. M., Tyagi, A. K., & Sreenath, N. (2021, January). The future with industry 4.0 at the core of society 5.0: Open issues, future opportunities and challenges. In *2021 International Conference on Computer Communication and Informatics (ICCCI)* (pp. 1–7). IEEE.

Narvaez Rojas, C., Alomia Peñafiel, G. A., Loaiza Buitrago, D. F., & Tavera Romero, C. A. (2021). Society 5.0: A Japanese concept for a superintelligent society. *Sustainability (Switzerland)*, 13(12). https://doi.org/10.3390/su13126567

Onday, O. (2019). Japan's society 5.0: Going beyond Industry 4.0. *Business and Economics Journal*, 10(2), 1–6.

Phuyal, S., Bista, D., & Bista, R. (2020). Challenges, opportunities and future directions of smart manufacturing: A state of art review. *Sustainable Futures*, 2, 100023.

Sarlan, A., Nadam, C., & Basri, S. (2014, November). Twitter sentiment analysis. In *Proceedings of the 6th International Conference on Information Technology and Multimedia* (pp. 212–216). IEEE.

Shiroishi, Y., Uchiyama, K., & Suzuki, N. (2018). Society 5.0: For human security and well-being. *Computer*, 51(7), 91–95.

Shiroishi, Y., Uchiyama, K., & Suzuki, N. (2019). Better actions for Society 5.0: Using AI for evidence-based policy making that keeps humans in the loop. *Computer*, 52(11), 73–78.

Tan, A. H. (1999, April). Text mining: The state of the art and the challenges. In *Proceedings of the Pakdd 1999 Workshop on Knowledge Disocovery from Advanced Databases* (Vol. 8, pp. 65–70). IEEE.

Tan, P. N., Steinbach, M., & Kumar, V. (2016). *Introduction to data mining*. Pearson Education India.

Yuko, H. (2017). Society 5.0: Aiming for a new human-centered society. Collaborative creation through global R&D open innovation for creating the future. *Hitachi Review*, 66(6), 8–13.

Zengin, Y., Naktiyok, S., Kaygın, E., Kavak, O., & Topçuoğlu, F. (2021). An investigation upon industry 4.0 and society 5.0 within the context of sustainable development goals. *Sustainability*, 13(5), 2682.

4

Barriers to the Adoption of Blockchain Technology in Indian Public Distribution System

Vernika Agarwal, Yadhukrishnan Vyppukkaran Krishnan,
Marina Marinelli, and Mukund Janardhanan

CONTENTS

4.1 Introduction

The public distribution system (PDS) of India is considered to be one of the world's largest distribution networks, which was implemented during World War II as a rationing system. The Indian PDS has undergone several improvements and policy changes since the establishment for keeping up the main goals such as storage, procurement and distribution of subsidized food grains and fuel. The scheme was revised in the 1990s to increase access to food grains for people living in hilly and remote areas as well to target the vulnerable (Tarozzi, 2005). The government introduced the Targeted Public Distribution System (TPDS) in 1997, focusing and connecting the poor and vulnerable via fair price shops (FPS) throughout the country (Chakraborty and Sarmah, 2019). TDPS helped in identifying 60 million households in the below poverty line (BPL); in 2000, the Antyodaya Anna Yojana scheme identified and included 10 million most priority households (Kishore and Chakrabarti, 2015). Under this scheme, priority households will receive 35 kg of food grains, and the others have access for 5 kg of the same. PDS is aimed to provide food security to poor houses. However, PDS has several flaws including inefficiency in recognizing intended beneficiaries, corruption, non-uniform transaction record and lack of integration among stakeholders' operations. Blockchain technology would provide the necessary impetus to move from demand-based process of service delivery to an eligibility-based system (Ølnes et al., 2017). Immutability of data in Blockchain and the transparency it provides would make it possible to provide the necessary trust required for automatically initiating service delivery by executing the smart contracts stored in the Blockchain through a consensus process (Golosova and Romanovs, 2018).

The policy-making agency of Government of India, NITI Aayog, has proposed using Blockchain to increase traceability in the PDS supply chain. While Blockchain is likely to solve many of the current PDS issues, as correctly perceived by NITI Aayog, it necessitates a detailed analysis of the complicated supply chain system and conceptualizes how technology can be efficiently utilized. Blockchain is mostly used in the creation of various cryptocurrencies, Ethereum, which was introduced in 2014, was the first significant step forward in Blockchain technology (Ølnes et al., 2017). Blockchain is an application layer that runs on top of the internet protocols and allows economic transactions between similar parties without the need for a third party (Tapscott, 2016). A collection of transactions registered to a log over a set period, resulting in the formation of a 'block' is a Blockchain scheme. Each transaction is recorded in a block as it occurs. Each block is connected to the blocks before and after it. These blocks are mathematically 'connected or chained' together using a hashing function; inside the Blockchain, it could consider a hash as a cryptographic fingerprint of data to jail it in framework (Laurence, 2019). As these blocks are paired in a sequence, they become permanent and cannot be modified by a single actor. Instead, they are managed and tested by automation and contributed governance protocols (Swan, 2017). Such change in operations of the current system could be benefited by improving the process efficiency by reducing the activities, minimum lead time of order handling and improved traceability of order (Martinez et al., 2019). The implementation of Blockchain in PDS will help to track down all the illegal activities and will help to improve the operating efficiency.

This study investigates the potential barriers in the implementation of Blockchain technology in the Indian PDS; classify them in line with their significance. This chapter uses expert input and systematic literature review to identify the obstacles to effective Blockchain adapting in the PDS, and the identified barriers are ranked using a novel multi-criteria decision-making technique of Best Worst Method (BWM).

The chapter is organized as follows. Section 4.2 provides a literature review with regard to Blockchain technology and PDS. This section also summarizes the identified set of barriers on the adoption of this technology. Section 4.3 provides the explanation and discussion of the adopted BWM technique in detail. Section 4.4 discusses the data analysis and results. Section 4.5 concludes the findings of this study and provides future research directions.

4.2 Literature Review

The successful integration of Blockchain technology in the Bitcoin and finance application led the idea towards integration of this technology to the future (Kshetri and Kshetri, 2018). Till date, Blockchain application in supply chain management is in early stages. With rising customer concern about food safety and security, companies are faced with the issue of implementing suitable safeguarding measures through proper application and integration of technology. This change is going to be beneficial to the Indian Public distribution system through digitization and proper resource management. Blockchain has lot of potentials to bring changes in the Indian PDS. However, because Indian PDSs are very complex, chaotic, semi-integrated, and include many intermediaries (Viswanadham and Samvedi, 2013), a proper integration platform is required, where technologies such as Blockchain might play an important role. Furthermore, as public knowledge grows, consumers are more worried about food safety and expect long-term compliance. Adoption of Blockchain technology is now necessary to ensure the above-mentioned worries of consumers in real-world contexts. However, Blockchain adoption is still in its infancy (Yadav et al., 2020).

Though Blockchain is likely to alleviate many of the existing PDS problems, as correctly assessed by NITI Aayog, it necessitates a comprehensive examination into the complex structure of the supply chain and conceptualizes how the technology may be efficiently implemented. A Blockchain implementation would create a massive amount and diversity of data. Although such data is ideal for analytics, employing it on-chain for decision support may result in increased delay. These concerns are addressed systematically by the suggested conceptual paradigm. It employs ration cards linked to a 12-digit Unique Identification Number (UID) to assign a unique identity to each household and proposes biometric technology to assure beneficiary identification (Singh et al., 2021). It can improve the government's access to PDS supply chain data for improved planning and distribution. Blockchain technology adoption in Indian PDS aims to track the necessary data during the supply chain of food grain distribution, which includes the minimum supportive price to farmers, price of commodity and delivered quantity during each stage of transportation to maintain a foolproof system (Yadav et al., 2021). The adoption will enable farmers to get quick payments for their supply to miller, and the millers will get paid instantly for the transportation to state warehouse. The merging of Blockchain will create a leakage-proof system by tracking down a time stamped entry log of commodities in warehouse at different stages of transportation. Integration of Blockchain technology in Indian PDS will improve the current issues faced by the beneficiaries that are present in the system (Kumar et al., 2021), which are discussed below:

- There are reports of consumers obtaining low-quality food grains at ration stores.
- Rogue dealers replace Food Corporation of India (FCI) supply with inferior stock and sell the good supply stock to private retailers.
- Illicit fair pricing store owners have been discovered creating a huge number of fraudulent cards to sell food grains on the open market.
- Due to the low pay, many dealers resort to fraud, illicit commodity diversions, holding, and black selling.
- Safe and nutritious food is inaccessible and unaffordable for many poor people, resulting in food insecurity.
- Delivery of PDS services has been very irregular and varied among states. With the recent introduction of Aadhar cards, the problem of identifying and delivering PDS services, as well as Direct Cash Transfers, has improved.
- The regional allocation and coverage of FPS are inadequate, and the primary goal of price stability of key goods has not been realized.
- There are no clear standards for determining whether a family is above or below the poverty line. This uncertainty creates a huge opportunity for corruption and blowback in PDS systems since those people who are supposed to profit are unable to do so.

4.2.1 Barriers in Adopting Blockchain in Indian PDS

Blockchain is a new technology with several potential uses. This technology, which supports cryptocurrencies, provides a distributed database of digital assets that is irreversible, decentralized, and transparent for use by enterprises in supply chains. However, not all businesses are well-suited to integrate Blockchain in the existing supply chain, owing to a lack of understanding of the technology's benefits. The most prominent hurdles to Blockchain adoption are the lack of corporate understanding and familiarity with Blockchain technology, and what it might bring for future supply chains and the complexity of Indian PDS makes it difficult for adoption.

A systematic literature research was carried out from the database of SCOPUS, Science Direct, Emerald, Springer, and Google scholar. The keywords used to sort out the literature were 'Blockchain', 'supply chain management', 'barriers', 'Indian PDS' and 'public distribution system'. Based on the research, four major kinds of impediments to Blockchain implementation in Indian PDS are identified: Intra-organizational, inter-organizational, system-related and external hurdles. Sub-barriers in each of these categories are summarized in Table 4.1.

a. **Intra-organizational barriers**: Intra-organizational obstacles are factors originating inside a company that cause managers to be hesitant

TABLE 4.1

Potential Barriers in the Implementation of Blockchain Technology

Major Criteria	Sub-Criteria	Description	References
Intra-organizational barriers (B1)	High implementation cost (B1.1)	Development cost, complexity	Soltani et al. (2005)
	Lack of knowledge and expertise (B1.2)	Lack of technical skills and technical developers	Soltani et al. (2005)
	Transparency versus privacy dilemma (B1.3)	Openness and privacy	Soltani et al. (2005)
	Lack of long-term planning (B1.4)	Complexity in predicting for the future	Soltani et al. (2005)
Inter-organizational barriers (B2)	Supply chain readiness (B2.1)	High inventory liability, the inability to match supply with demand in real time and a lack of alternate supply sources	Treiblmaier et al. (2021)
	Inaccurate inputs (B2.2)	Manipulations or mistakes makes erroneous data into the system	Treiblmaier et al. (2021)
	Variations in standards (B2.3)	Lack of standards in information traceability and data format	Treiblmaier et al. (2021)
	Lack of trust (B2.4)	Lack of trust between various channel partners	OWN
System-related barriers (B3)	Technology complexity and data access (B3.1)	Lack of technological access to obtain real-time information in a PDS	Abeyratne and Monfared (2016)
	Scalability (B3.2)	Updating process becomes slower, and latency becomes more of a concern	Abeyratne and Monfared (2016)
	Smart contract designing (B3.3)	The capacity to self-execute based on specified circumstances	Abeyratne and Monfared (2016)
	Lack of real-time access (B3.4)	The data is often obsolete	Abeyratne and Monfared (2016)
External barriers (B4)	Government support and funding (B4.1)	Lack of government assistance in the form of funding or supporting policies inhibits companies from contemplating using the technology	Saberi et al. (2019)
	Privacy leakage (B4.2)	Blockchain could not guarantee transactional privacy, negative impact on user privacy protection	Treiblmaier et al. (2021)
	Weak grievance redressal (B4.3)	Lack of new age technologies in grievance redressal systems, mostly the traditional approaches are used.	OWN
	Lack of consistent policies (B4.4)	The changing government policies regarding impact the supply chain of the PDS	OWN

to embrace Blockchain. These obstacles are the result of organizational internal activity. The backing of top management is critical to the effective adoption of any supply chain strategy. However, some managers lack the long-term commitment and support required to implement new technologies while adhering to sustainable ideas (Soltani et al., 2005).

b. **Inter-organizational barriers**: These barriers occur when there are collaboration issues between/involving two or more organizations. There are several inter-organizational obstacles impeding Blockchain deployment. This category primarily detects and establishes relationship obstacles between supply chain participants. Supply chain management is primarily concerned with managing connections among partners to produce value for stakeholders (Treiblmaier et al., 2021).

c. **System-related barriers**: New IT tools are required to deploy Blockchain technology and process the information related to the operation of PDS. This can be difficult for some PDS members (Abeyratne and Monfared, 2016). To take advantage of the potential for value savings in the PDS, all members must have access to the necessary information. As a result, the lack of technological access to obtain real-time information in a PDS is a barrier to using Blockchain technology.

d. **External barriers**: External barriers are obstacles posed by external stakeholders like as institutions or the government that are not immediately benefited by the operations (Saberi et al., 2019; Treiblmaier et al., 2021). Participants in a complicated supply chain, such as food, may be in various areas, putting them under distinct constraints and laws (Janvier-James, 2012).

4.2.2 Research Gaps

Based on the literature study, it was evident that there is no work reported that focusses on the barriers of adoption of Blockchain in Indian PDS and barrier ranking using MCDM approach. Identifying the adoption barriers will provide a better understanding of the underlying issues and processes for effective Blockchain use in Indian PDS. Further modelling of these obstacles would aid in understanding their current interrelationship. Obtaining the causal link between the identified barriers would aid in determining the severity of the relationship between the identified barriers. Finally, grouping these barriers would aid in obtaining detailed driving and the force of each barrier's dependency. Such data would assist the government, relevant policy-making authorities, organizations, and Indian PDS stakeholders in developing a viable plan for the successful implementation of Blockchain in Indian PDS.

4.3 Research Methodology

To formulate a hierarchical list of barriers, we have applied the BWM, aligning with the objective of this study, and we have consulted five stakeholders, having 10–15 years of experience to give their inputs. BWM has helped us in generating weights of these barriers using only two vectors which makes this method more relevant in comparison to other MCDM techniques. Our technique only requires 'best-to-others' and 'others-to-best' vectors, thereby reducing the complexity and decision-making time. Let the chosen challenge sets be $SV = \{SV_1, SV_2, ..., SV_9\}$. The BWM model makes decision of prioritizing the challenges after undergoing the following steps:

Step 1: Selecting the most and least critical barrier.

Initially, the most and least critical barrier are chosen based on the input of each stakeholder.

Step 2: Determining the most critical barrier over decision set.

This step involves evaluating the most critical barrier based on the pairwise comparison made using scale of 1–9.

Formula for calculating 'best-to-others' resulting vector is as follows:

$$SV_B = \left(sv_{B1}, ..., sv_{B9}\right)$$

where sv_{Bi} gives preference to the most critical barrier over ith challenge and $sv_{BB} = 1$.

Step 3: Calculating the preference of the least critical barrier over decision set.

This step utilizes the pairwise comparison to validate the preference of other barrier over the least critical barrier, again using the scale of 1–9.

Formula for calculating "worst-to-others" resulting vector is as follows:

$$SV_W = \left(sv_{1W}, ..., sv_{9W}\right)^T$$

where sv_{Wi} gives the preference to the least critical barrier over ith challenge and $sv_{WW} = 1$.

Step 4: Calculating the optimum weights of barrier.

This step aims at calculating the optimum weight vector $\left(z_1^*, ..., z_9^*\right)$ of the barrier.

The optimum weight of the ith challenge will meet the below mentioned requirements:

$$\frac{z_B^*}{z_i^*} = sv_{Bi} \quad \text{and} \quad \frac{z_i^*}{z_W^*} = sv_{iw}$$

.

To satisfy the condition, the Maximum Absolute Difference $\left|\dfrac{z_B^*}{z_i^*} - \text{sv}_{Bi}\right|$ and $\left|\dfrac{z_i^*}{z_W^*} - \text{sv}_{iw}\right|$ should be minimized for every barriers. The optimum weights for barriers can be attained by applying the programming problem as follows (Rezaei, 2015):

$$\min_i \max \left\{ \left|\frac{z_B^*}{z_i^*} - \text{sv}_{Bi}\right| , \left|\frac{z_i^*}{z_W^*} - \text{sv}_{iw}\right| \right\}$$

Subject to

$$\sum_I z_i = 1 \tag{P1}$$

$$z_i \geq 0 \qquad \forall i = 1, 2, ..., 9$$

Problem (P1) is equivalent to the following linear programming formulation (P2):

$$\min \phi$$

Subject to

$$|z_B - \text{sv}_{Bi} z_i| \leq \phi \qquad \forall i = 1, 2, ..., 9;$$

$$|z_i - \text{sv}_W z_W| \leq \phi \qquad \forall i = 1, 2, ..., 9;$$

$$\sum_I z_i = 1 \tag{P2}$$

$$z_i \geq 0 \qquad \forall i = 1, 2, ..., 9;$$

Solution of this problem gives us the ratio of consistency ϕ^* and the optimum weights as $(z_1^*, ..., z_9^*)$.

The closer is the ratio of consistency ϕ^* to the zero value, the more is the consistency of the system.

Step 5: Checking the solution's consistency.

The closer is the ratio of consistency ϕ^* to the zero value, the more is the consistency of the system of comparison provided by the experts. Solution's consistency can verified by the calculation of the ratio of consistency:

$$\text{Consistency Ratio} = \frac{\phi^*}{\text{Consistency Index}}$$

Table 4.2 is used to get the value of the consistency index (Rezaei, 2015).

TABLE 4.2

Consistency Index Table for BWM

vBi	1	2	3	4	5	6	7	8	9
Consistency index (max)	0.00	0.44	1.00	1.63	2.30	3.00	3.73	4.47	5.23

TABLE 4.3

Rating of the Stakeholder for Best-to-Others and Others-to-Best Vectors for Intra-Organizational Barriers

	BO					OW				
	DM1	DM2	DM3	DM4	DM5	DM1	DM2	DM3	DM4	DM5
B1.1	1	3	1	9	7	9	7	9	1	3
B1.2	4	1	5	1	9	5	9	4	9	1
B1.3	6	5	9	2	1	4	3	1	8	9
B1.4	9	9	3	6	2	1	1	7	4	6

Values of the ratio of consistency close to zero depict more consistency, whereas the values closer to one depict less consistency. The ratio of consistency which is not 0 signifies that the pairwise comparison matrix is partially consistent, and we might have multiple optimality.

Step 6: Steps 1–5 are repeated for each stakeholder. To get the final weights, the average of all the weights is considered.

4.4 Data Analysis

The flaws such as corruption, non-uniform transaction record and inefficient database in the existing Indian PDS have paved the way for the emerging technology such as Blockchain. These new age technologies have the potential to change the way we live, work and relate to one another including the operations of PDS (Mishra and Maheshwari, 2021). This section presents the analysis of the data to verify the anticipated framework. Following the BWM steps as mentioned in research methodology section, the barriers have been ranked based on their criticality. Since the BWM needs only a few variables, it becomes easier for the decision makers to choose the criteria for the least critical and the best critical barrier. Table 4.3 demonstrates the rating of the stakeholder for best-to-others and others-to-best vectors for Intra-Organizational Barriers (B1).

Similarly, the inputs of the stakeholders were taken for other categories of barriers. Problem P2 of Linear Programming is used in Step 4 to determine the weights. The ratio of consistency ϕ^* and the ideal weight can be found out

TABLE 4.4

Weights of the Barriers and Their Rankings

Major Criteria	Local Weights	Sub-Criteria	Local Weights	Global Weights	Rank
B1	0.2328	B1.1	0.3160	0.0736	4
		B1.2	0.2978	0.0693	5
		B1.3	0.2351	0.0547	7
		B1.4	0.1511	0.0352	10
B2	0.5769	B2.1	0.3259	0.1880	2
		B2.2	0.2450	0.1414	3
		B2.3	0.3144	0.1814	1
		B2.4	0.1146	0.0661	6
B3	0.1397	B3.1	0.3577	0.0500	8
		B3.2	0.2429	0.0339	11
		B3.3	0.0838	0.0117	14
		B3.4	0.3156	0.0441	9
B4	0.0506	B4.1	0.3102	0.0157	13
		B4.2	0.2140	0.0108	15
		B4.3	0.4051	0.0205	12
		B4.4	0.0707	0.0036	16

by solving P2. The result shows that the consistency is within range for all the challenges. Problem P2 is used to calculate the optimum weights for the challenges. Now after calculation of the average of these weights, these barriers can be appropriately ranked, as shown in Table 4.4.

In this chapter, we aim in calculating the weights of these barriers based on their criticality levels. The challenges with more weight tend to be the ones with higher critical levels and require immediate attention. Based on the ranking done using the BWM technique, it is evident that among the barriers, Variations in standards (B2.3) are ranked first because it has the maximum weightage of 0.1814. Supply chain readiness (B2.1) is ranked second with the weightage of 0.1880. It can be observed that the highest weights are for inter-organizational barriers; these occur at the base level when the collaboration between the organizations is lacking. Supply chain management is primarily concerned with managing connections among partners to produce value for stakeholders.

4.5 Conclusion

The PDS of India is the world's largest distribution network, covering the length and breadth of the country. The Indian PDS has undergone several improvements and policy changes since the establishment for keeping up the main goals such as storage, procurement and distribution of subsidized food grains and fuel. The new age technologies, like Blockchain, are now introduced to enhance traceability in the PDS supply chain. The implementation of Blockchain is, however, hindered by a number of internal and external barriers. This chapter aims in understanding these impediments for Blockchain adoption to Indian PDS with the help of industry experts' opinion. The prioritization of identified barriers is done using BWM. Variations in standards and supply chain readiness are two crucial barriers identified by the study. Delivery of PDS services has been very irregular and varied among states. In addition, high inventory liability, the inability to match supply with demand in real time and a lack of alternate supply sources are delaying the supply chains to reach the target audience. These hurdles obstruct and influence firms' decisions to develop a Blockchain-enabled supply chain in the process of public distribution, and other barriers operate as secondary and related factors in the adoption process.

References

Abeyratne, S. A. and R. P. Monfared (2016). "Blockchain ready manufacturing supply chain using distributed ledger". *International Journal of Research in Engineering and Technology* 5(9): 1–10.

Chakraborty, S. and S. Sarmah (2019). "India 2025: The public distribution system and national food security act 2013". *Development in Practice* 29(2): 230–249.

Golosova, J. and A. Romanovs (2018). The advantages and disadvantages of the blockchain technology. In *2018 IEEE 6th workshop on advances in information, electronic and electrical engineering (AIEEE)*. IEEE.

Janvier-James, A. M. (2012). "A new introduction to supply chains and supply chain management: Definitions and theories perspective". *International Business Research* 5(1): 194–207.

Kishore, A. and S. Chakrabarti (2015). "Is more inclusive more effective? The 'new style' public distribution system in India". *Food Policy* 55: 117–130.

Kshetri, N. and N. Kshetri (2018). "The Indian blockchain landscape: Regulations and policy measures". *Asian Res. Policy* 9(2): 56–71.

Kumar, S., R. D. Raut, M. M. Queiroz and B. E. Narkhede (2021). "Mapping the barriers of AI implementations in the public distribution system: The Indian experience". *Technology in Society* 67: 101737.

Laurence, T. (2019). *Blockchain for dummies*. John Wiley & Sons.

Martinez, V., M. Zhao, C. Blujdea, X. Han, A. Neely and P. Albores (2019). "Blockchain-driven customer order management". *International Journal of Operations & Production Management* 39(6/7/8): 993–1022.

Mishra, H. and P. Maheshwari (2021). "Blockchain in Indian Public Distribution System: A conceptual framework to prevent leakage of the supplies and its enablers and disablers". *Journal of Global Operations and Strategic Sourcing.* 14(2): 312–335

Ølnes, S., J. Ubacht and M. Janssen (2017). "Blockchain in government: Benefits and implications of distributed ledger technology for information sharing". *Elsevier* 34: 355–364.

Rezaei, J. (2015). "Best-worst multi-criteria decision-making method". *Omega* 53: 49–57.

Saberi, S., M. Kouhizadeh, J. Sarkis and L. Shen (2019). "Blockchain technology and its relationships to sustainable supply chain management". *International Journal of Production Research* 57(7): 2117–2135.

Singh, S. K., M. Jenamani, D. Dasgupta and S. Das (2021). "A conceptual model for Indian public distribution system using consortium blockchain with on-chain and off-chain trusted data". *Information Technology for Development* 27(3): 499–523.

Soltani, E., P.-C. Lai and N. S. Gharneh (2005). "Breaking through barriers to TQM effectiveness: Lack of commitment of upper-level management". *Total Quality Management and Business Excellence* 16(8–9): 1009–1021.

Swan, M. (2017). "Anticipating the economic benefits of blockchain". *Technology Innovation Management Review* 7(10): 6–13.

Tapscott, D. (2016). *How the blockchain is changing money and business.* TED Summit.

Tarozzi, A. (2005). "The Indian Public Distribution System as provider of food security: Evidence from child nutrition in Andhra Pradesh". *European Economic Review* 49(5): 1305–1330.

Treiblmaier, H., A. Rejeb, R. van Hoek and M. Lacity (2021). "Intra- and interorganizational barriers to blockchain adoption: A general assessment and coping strategies in the agrifood industry". *Logistics* 5(4): 87.

Viswanadham, N. and A. Samvedi (2013). "Supplier selection based on supply chain ecosystem, performance and risk criteria". *International Journal of Production Research* 51(21): 6484–6498.

Yadav, V. S., A. Singh, R. D. Raut and N. Cheikhrouhou (2021). "Blockchain drivers to achieve sustainable food security in the Indian context". *Annals of Operations Research*: 1–39.

Yadav, V. S., A. R. Singh, R. D. Raut and U. H. Govindarajan (2020). "Blockchain technology adoption barriers in the Indian agricultural supply chain: An integrated approach". *Resources, Conservation and Recycling* 161: 104877.

5

Exploring the Driving and Restraining Forces to Fostering Blockchain in Sustainable Supply Chain in the Context of COVID-19 Pandemic

S. Ramabalan, K. Nagalakshmi, R. Anand Babu,
K. Raju, R. Lavanya, and R. Karthi

CONTENTS

5.1 Introduction

The pandemic has now reached the entire humanity on the earth triggering severe health crises and financial uncertainty globally. It has led to piling higher mortality in patients with comorbidities and engendered outpouring shocks on both demand-side and supply-side of global trade. That is, it marks a huge dent in the international economy and financial markets with disruptions of numerous industries such as automotive, aviation, oil, construction, food, healthcare, tourism and transport, insurance, telecommunications,

DOI: 10.1201/9781003177432-7

agriculture, retail industry, and supply networks imposing governments and organizations to lockdown operations on a global scale (Huang et al., 2020).

Due to this viral pandemic, an intact supply chain, from raw material providers to delivery docks of consumers, is broken up indeterminately. Personal protective equipment, ventilators, masks, testing kits, and even commodities required for daily care have been scarce as lockdown and social distancing strategies have kicked in. Also, the pandemic cuts international shipment of raw materials and critical accessories by about $228 billion due to disruptions of the global SC network (Solleder and Velasquez, 2020). According to a survey reported by Capgemini Research Institute, more than 80% of industries are being undesirably affected by the pandemic, and most of them have contended with major challenges as illustrated in Figure 5.1 (Capgemini Research Institute, 2020). These include the difficulties in end-to-end monitoring of the SC (72%), the deficit of critical supplies (74%), deferred deliveries and longer lead times (74%), obstacles in regulating manufacturing capacity against sharp spikes, declines in demands (69%), and complications in development amongst volatile demand (68%), etc. All the enterprises such as discrete manufacturing, retail, life sciences, and consumer products experienced similar challenges across their supply networks.

During the pandemic, applying and executing key SC processes in a secured, trusted, effective, universally manageable, and traceable way is a perplexing task owing to its fragile nature, which is susceptible to redundant efforts and systemic risks that can cause adversarial effects on customer well-being and safety. Recently, BCT has acquired worldwide

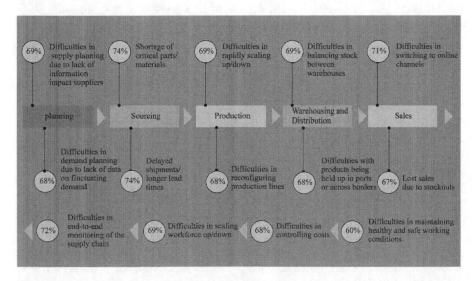

FIGURE 5.1
Percentage of organizations that encountered considerable challenges across the supply chain during the pandemic. (Capgemini Research Institute, 2020.)

hyperawareness with the capacity to reshape SC administration and sustainability accomplishments. The few applied and ongoing applications show how corporates strive to apply BCT into their SC processes, as in the case of Walmart (Kshetri, 2018), Provenance (Baker, 2012), Maersk (Popper & Lohr, 2017), and of late in Mongolia for improving the viability of cashmere (Huang, 2019). Several corporates utilize BCT for curtailing forged goods (Fernandez-Carames & Fraga-Lamas, 2018) and food safety (Casey &Wong, 2017). The above-mentioned corporates exploit SC procedures for achieving security, safety, and globally traceable products in their trades by improving sustainability in their network. In spite of the several BCT reimbursements for increasing sustainability in a supply chain, the applications adopting BCT for sustainability are meager, whereas organizations still grapple to implement all-inclusive characteristics of sustainable SC. Moreover, the financial support for BCT, with some exemptions, is declining. This statistic will be another stimulus for this work that encourages us to analyze significant forces (driving/restraining forces), which accelerate/decelerate the fostering of BCT in sustainable SC networks.

To date, SC sustainability becomes a key enabler for consumer demand and trustworthiness. It is defined as a trade-off between business, social, and environmental conditions, also called the triple-bottom-line (Seuring et al., 2008). There are regulatory, viable, and social causes for supporting reliable SC. Customers attempt to attest to their items and need a portal to access their information and goods. This condition has placed a burden on vendors to provide sustainable SCs on a local and global scale. At present, there are several auditing and data sustainability certification schemes are available for formulating any evidence related to the appraising of such data. Conversely, these systems include controlled databanks which indicate that their validity and credibility can be interrogated. BCT aids these sustainability authorizations that flow deep into the SC network.

Blockchain offers distributed records comprising data on auditable interactions and the digital cryptocurrency (e.g., Bitcoin). The network starts with an initial block (i.e., genesis block). When a new block is included in the chains, it is connected to its ancestors using a cryptographic pointer (Alladi et al., 2019). The cryptographic sealing is employed to encrypt wallets or places on the block where value or work is securely stored. These tamper-proof records are processed by a public timestamping server through a peer-to-peer (P2P) network. Thus, decentralized consensus, safe, certified, visible, and transparent data are all vital properties of BCT. Hence, this technology can be deployed toward digital transformation in the supply network for circumventing potential risk factors and challenges (Dutta et al., 2020).

The performance of SC in monitoring and tracking is increased by integrating decentralized logs of BCT (Rejeb et al., 2019). It provides numerous possible solutions to improve scalability, the speed, and visibility of SC, circumventing fake merchandises trades, handling product recalls, expiration, deficiencies while improving batching, distribution, and inventory management. It makes business dealings more secure, transparent, and tamper-proof, which can have a colossal impact on the efficiency of the sustainable SC. Besides, it will improve the enactment of SC further with live tracking in improved and complete visibility. Albeit, the pandemic has made an immense disruption in supply network worldwide BCT fetches some useful insights into how SC can be made more robust and how industries find out sustainable solutions to handle disruptions or pandemics.

In spite of the potential benefits of BCT, the speed of fostering in sustainable SC management system has been slowed down. Most of the ongoing applications found in the literature are halted at the experimental phase. In this chapter, we explore how fostering BCT in sustainable SC processes with several environmental, organizational, and technological promises has been delayed. Therefore, it is essential to understand the potential driving and restraining forces that organizations might experience during the implementation of BCT in their trades. Hence, this chapter will carry out a systematic study of how BCT fits in the modern SC management system and explore possible challenges with its application.

5.2 Blockchain Technology

BCT emanated into the limelight as a cryptographically protected and distributed ledger comprising information on auditable interactions and digital cryptocurrency in 2008 (Nakamoto, 2008). Recently, beyond its best-known application of Bitcoin, Blockchain has gained prevalent attention and interest from several industries and communities including government (Clavin et al., 2020), healthcare (Abu-elezz et al., 2020), finance (Pal et al., 2021), supply chain (Narayan and Tidström, 2020), etc.

According to a recent forecast by Gartner, the global economy amplified by BCT will be reached about $176 billion by 2025, then outpouring to surpass $3,100 billion by 2030 (Gartner, 2019). Figure 5.2 depicts the top five predictions of BCT for 2030 (Mitselmakher, 2018). According to these estimates, by 2030, most governments will espouse or create different virtual currencies in most (perhaps all) of their commercial activities. In the future BCT era, trillion-dollar tokens will act a vital role in improving the flow of the digital economy. BCT enables frictionless transactions of tokens and other resources. By 2030 or even earlier, all individuals and their virtual/physical resources will have Blockchain identities. BCT will help enhance systems for handling those identities locally as well as globally using several identity solutions (Mitselmakher, 2018). By 2030, noteworthy progress in the quality of life will be attributable to the emerging BCT network. It is also expected that most of the world trade will be carried out through BCT technology.

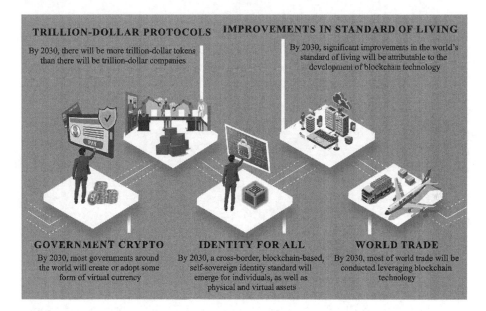

FIGURE 5.2
Top five predictions.

5.2.1 Components of Data Flow in BCT

In order to the meet diverse requirements of different industries and consumers, numerous forms of BCT networks are developed, and each comprises several unique characteristics; however, a rudimentary basis remains unchanged for all. Here, we consider Bitcoin, the primary and the most popular BCT network, to demonstrate the major elements of conventional data sharing mechanism in a BCT network:

- Node: Node is a user who partakes in interaction on the BCT. Fundamentally, it keeps a comprehensive and immutable record that contains all the past communications. It acts as a keystone to save a complete replica of the secured record in every node. Then, a node participates in the communication by disseminating transactional information and allowing miners to create and validate each block.

- Block: It is a data structure that is deployed to gather a group of secured interactions by including a hash value to guarantee the credibility of the transaction. Block is an indispensable element and is employed in all BCT-based applications.

- Digital wallet: It is a protected warehouse for a customer to hoard the public and private keys. It communicates with the BCT; therefore, a customer can transmit and receive virtual coinage as well as track their transactions.

- Miner: It is a distinct consumer in the BCT network, gathers and authenticates all interactions, and generates new blocks. It contests with other miners to resolve a riddle, usually called a proof-of-work (PoW) consensus mechanism. The first to solve the riddle includes a new block to the network and gets a definite remuneration (e.g., some Bitcoins). After including a new block, all the nodes harmonize their local replica, guaranteeing their record is the latest. A mining procedure or miner is employed for authentication in several public BCT, while authentication is restricted by a consensus in a private BCT network.

- Consensus: It is one of the most vital elements of BCT and represents a contract among nodes that provides data. A network is rationalized through an organized consensus algorithm to guarantee that blocks and transactions are organized in order properly, to ensure the stability and reliability of the records to increase reliance among nodes (stakeholders). Furthermore, a consensus protocol can support the distributed network consistently to make an appropriate decision (Gupta, 2018). Predominant consensus protocol contains proof-of-stake, proof-of-authority matched, delegated proof-of-stake, Byzantine fault tolerance, proof-of-work, and proof-of-elapsed time (Baliga, 2017).

Each block comprises the hash code of the preceding block, a timestamp, and a group of verified interactions (Murthy et al., 2020). The genesis block (i.e., initial block) has no previous block, as shown in Figure 5.3. The transactional data in BCT are permanent and cannot be altered once they are legitimately authorized by a consensus-based process and recorded into the ledger. The data appraisal is instantly distributed across the chain. Hence, it enables customers to audit and verify their communications autonomously and transparently. The distributed and transparent nature of BCT enables users to save and track transactional data efficiently. These features can also eliminate the double-spending problem (Uddin et al., 2021). Double-spending is a common issue in a cryptocurrency system where the same single token can be expended more than once. Also, BCT is a trustless computing environment in which the validity of transactional data is guaranteed without any third-party authorization (Uddin et al., 2021).

- Figure 5.3 describes the overall structure of BCT. The blocks contain a body and a header. The header consists of the following six fields:
- Version (4 bytes): This field denotes the consensus algorithm to be used in the current application.
- Previous hash code (32 bytes): It represents the hash code of the preceding block. Without this field, there is no link and chronology among logs.

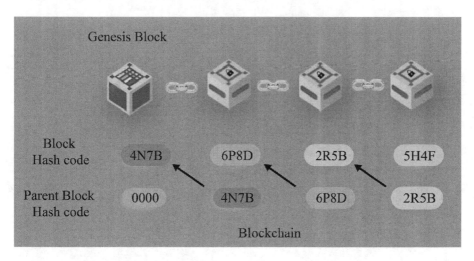

FIGURE 5.3
Overview of BCT technology.

- Merkel root: This field is employed to identify each interaction. This field is represented as a binary tree of the hash values. It is used in Bitcoin to make the data blocks tamper-proof.
- Difficulty target (4 bytes): This field calculates the target threshold of hash value to identify the legal block.
- Nonce (4 bytes): Nonce is a random value used for secure transactions. In general, it starts with zero and increases gradually for each hash calculation.
- Timestamp (4 bytes): It is given in seconds since 1/1/1970. It is used to ensure block integrity by monitoring the generation and modification time of a record securely. At this point, security denotes that nobody (not even the proprietor of the record) can modify the block once it has been registered provided that the data reliability is not ever negotiated.

The body of the block contains recognized and validated transactions. The transaction counter is used to count all the transactions as shown in Figure 5.4. The state of the block denotes who transferred which data to whom at a particular time. A recognized transaction among two peers only befalls when it is involved in a block then it is certified and linked to the chain. Hence, the log must be accessible publicly. This demonstrates the inevitability of P2P networks in BCT (Litke et al., 2019). The working principle of BCT when a customer sends transactional data (e.g., cryptocurrency) to another customer is illustrated by the following steps as given in Figure 5.5.

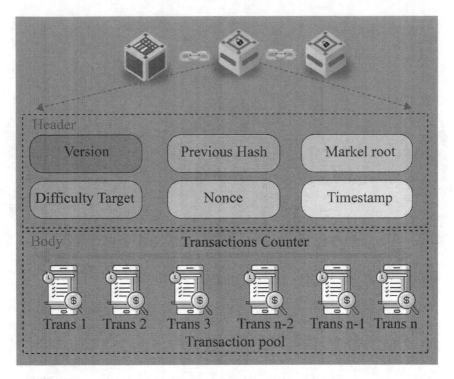

FIGURE 5.4
General structure of a block.

1. Consider a stakeholder needs to send a definite amount of virtual currencies to another stakeholder. The sender starts a transaction. Participants can frequently utilize their personal assistant devices like mobile phones or laptops for creating business dealings. The contents are authorized with the sender's private key and if required, the transactions and information are encoded with the public key of the receiver.

2. The sender directs the transaction through a P2P network containing nodes (usually high-speed computers). Different consensus algorithms are applied on this network to check the dependability of the block, circumvent fraud-related deeds including double-spending threats, and guarantee smooth information sharing.

3. These nodes duplicate the transaction and disseminate it to every party in the network. Hence, the nodes have a file of interactions (ledger). Figure 5.4 illustrates the configuration of a basic block.

4. Each participant can add blocks to the prevailing system of verified blocks only if a target hash code is formed by resolving intricate PoW. This is known as consensus protocol and it differs with respect to processing overhead and running time.

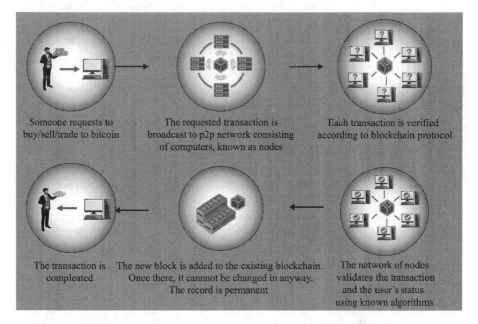

Someone requests to buy/sell/trade to bitcoin

The requested transaction is broadcast to p2p network consisting of computers, known as nodes

Each transaction is verified according to blockchain protocol

The transaction is compleated

The new block is added to the existing blockchain. Once there, it cannnot be changed in anyway. The record is permanent

The network of nodes validates the transaction and the user's status using known algorithms

FIGURE 5.5
Working mechanism of BCT.

5. Now the receiver participant can get the transaction details from the verified block using its private key.

5.2.2 Public versus Private Blockchain

Based on the decentralized structures and the technologies bundled in that, the BCT platforms can be characterized as either a public or private network. In public BCT, an open network is available for each stakeholder and enables them to participate in consensus validation. Hence, it is also known as permissionless or trustless networks. They are completely distributed across anonymous parties. Within this kind of system, confidence among customers is inadequate or missing. To evade this deficiency, miners are used to authorize transactions in this network. Conversely, private BCT is a collection of recognized customers working under a consensus mechanism, to increase organizational reliance. To connect these networks, a new user requests authorization from the existing participants or a delegated customer; therefore, it is also known as a permissioned BCT. This type of network enables confidence among consumers and does not need expensive miners. More effective consensus mechanisms (e.g., application of Byzantine fault-tolerant protocol) are used to authorize information, enhance system output, and decrease the latency of dealings.

5.2.3 Key Characteristics

The following key characteristics make BCT superior to any other centralized database management system:

- Decentralization: In BCT, the transactions are managed and validated using decentralized records (i.e., individual participant datasets). Hence, a single-point failure problem is not possible. Moreover, a BCT is not governed by any selected person, authority, or organization, and the requirement for reliable mediators to check transactions is removed. For instance, in the cryptocurrency system, it is not mandatory for any reliable third party such as Bank. All participants can cooperate on all aspects for verifying and connecting blocks in the chain. It adopts a strong platform that increases the dependability, availability, and quality of services and data. Consequently, complete transparency can be preserved among numerous organizations with intact transaction streams, products, and information in an efficient and rapid way.

- Persistency: It is not possible to roll back or discard a registered transaction. However, illegal transactions are identified instantly. After checking transactions by the consensus protocol, all the data are recorded. Subsequently, the captured information cannot be altered.

- Anonymity: The user exploits virtual identity code to interact with the BCT network. The process of virtual identification does not expose the actual user identity. Therefore, this characteristic brings numerous privacy and security issues in BCT-based applications (Kosba et al., 2016).

- Auditability: This characteristic represents the secure connection between every block and the previous one. It makes transactional data monitored and verified easily.

- Versatile value sharing: BCT enables an effective and secure network for storing the transactions of assets, the provenance of goods and services, intellectual property rights, Bitcoin exchange, etc.

- Improved cyber security: BCT exploits digital signature algorithms and asymmetric cryptography to guarantee individual identity as well as data security.

5.3 Sustainable Supply Chain Today

The supply chain includes the complete flow, such as the real-time and related information flow of basic material, goods, data, and currency. It acts

FIGURE 5.6
Supply chain and operations.

a distinctive and vital role in industries/trades and regulates the enactment of firms. SC achieves data flow from raw material sourcing to production, delivery, and logistics. Therefore, it impacts the capital investments,
speed-to-market, price of goods, and service insights in trades. SC assimi-
lates a collection of segmented and often physically distinct endeavors into
a hybrid network to provide value to the customer. Figure 5.6 depicts the
key operations of a conventional SC management system. The basic building
blocks of the SC management system are forecasting and planning, procurement, inventory management, manufacturing and scheduling, distribution
and delivery, and customer service.

SC prediction combines data from previous trades with understandings
and insights into increasing new markets, demand, cargo booking, stock
inventory, and budget planning. Planning is the procedure of correctly planning the flow of materials or products from the raw material stage to delivery
docks of consumers. This phase contains the planning of supply, demand,
manufacturing, delivery, processes, and sales. Procurement is the effort of an
organization to determine and acquire materials. This phase contains sealing contracts, finding supplies, handling contractors, and making invoice
settlements.

As part of the SC operations, inventory management includes governing
and managing procurements from suppliers and consumers, preserving
the storage of stock, monitoring the goods for sale, and order fulfillment.
Manufacturing scheduling is the operation of allocating diverse resources,
raw materials, and processes to various goods. It makes manufacturing as
effective and profitable as possible by distributing goods on time. Distribution
and delivery include the efforts taken to transfer and stock merchandise from
the supplier point to a customer point in the SC. The entire obstacles and the
burdens of the users are identified by customer service, and this information
can support to enhance the SC processes.

To date, the range of trades has expanded, the number of physical locations associated with the manufacturing operation has developed, and
product portfolios have expanded. Moreover, international SCs have been
rigorously suffered from the widespread disruptive pandemic (Ivanov &
Das, 2020). The interruptions of SCs are growing at a shocking rate with the

government-mandated lockdowns, imposed quarantines, entry bans, and strict travel restrictions. Accordingly, the SC has developed from a conventional supplier's network and manufacturers to a huge environment made of different goods that move through several parties and need collaboration among participants. Besides, owing to the quick development of e-commerce, the need for better-quality goods traceability and visibility across the development lifecycle has been lower. Conversely, the ineffectiveness of data communication in present SC vividly impacts the actions of manufacturers and retailers. For instance, gaps between information gathered by industrial units and by vendors make it difficult to track goods information and provide custom-made goods.

To increase the performance of SC and overcome the above-mentioned hitches, organizations have discovered new approaches that enable effective cooperation within and among diverse corporates. Amongst these technologies, BCT enables an auspicious future and enables the SC to offer improved transparency, visibility, and acuity of businesses during the course of the processes (Pilkington, 2016). Though BCT-based implementations in the SC are still in their initial phases, we trust this technology will considerably reshape the SC network (Kshetri, 2018). Experts predict that BCT can help the SC system gain one-third of enhancement in most of its general operations (Gunasekaran & Arha, 2019). As mentioned earlier, BCT delivers a method to generate tamper-proof transaction ledgers and information. Transactional information is irreversible since it cannot be altered once they are disseminated, established, and authorized by consensus protocol and hoarded in the blocks. By excluding mediators to realize confidence amid all participants, effectiveness increases, and the budget is decreased for the complete SC network.

In spite of the wide-ranging acceptance that BCT enables quicker, more easily verifiable transactions and provides immutable data communication amongst SC stakeholders, it will take time for this technology to be fostered and to transform the sustainable SC system. At present, most of the BCT applications are theoretical expositions, and experiential proof on the application of it is inadequate. Additionally, some works have been performed on the challenges of implementing BCT in sustainable SC, including scalability, technical proficiency, organizational readiness, and interoperability with prevailing structures. Hence, this work will provide an analysis of how BCT fits in the sustainable SC system and confer possible barriers with its application.

5.3.1 Challenges in SC within a Pandemic Context

Changing consumer needs, physically alienated processes, challenges from the competitive market, and fostering innovative corporate models make the existing SC an extremely intricate structure. Recently, personal digital assistants and new business models have considerably transformed our daily lives, particularly our shopping experience. There is a growing demand for personalized goods, competent yet simplified spending experiences, and transparency about the cost and attribution of products. These demands fetch new openings to transactions but enact noteworthy challenges to present SCs. The conventional SCs grapple to enable data visibility for the end-to-end flow, monitor products from raw material to end-user, and enhance demand management, all of which are extremely intricate. Additionally, the technology of traditional SC is failing to deliver acceptable crisis management, minimize outlays, or adapt to fluctuating consumer demands. We review the major barriers in present SCs:

- Lack of transparency: SC transparency denotes the level to which the participants access to and own a shared knowledge of sufficient and correct product information. The SC transparency system develops confidence among participants and ensures the reliability of goods and related information. Conversely, the distributed datasets in existing SC systems provide nominal transparency, and most of the valuable information is missing when goods and data are transported from one participant to another. Besides, there are problems with unreliable communication, reliance on paper documents, and poor compatibility. These serious issues continue even with years of substantial research contribution. The Chipotle Mexican Grill outlets is a sad and significant case of in what way the present SC network is ineffective at, and probably unable to contribute transparency during the course of the complete lifespan of goods delivery (Kshetri, 2018).

- Limited traceability: Traceability is a vital issue for SC, particularly regarding customer service, in predicting and scheduling commercial activities. Conversely, it is very challenging to implement a centralized model in a networked environment, mainly where confidence amongst stakeholders is inadequate. As an alternative, there are some distributed networks among participants that contain different databases that hamper commodity tracing during the course of the complete SC system (Maouchi et al., 2018).

- Participant's cynicism: Trust is an indispensable issue in SC management, and an efficient SC system must be created on a strong basis of it. Still, cynicism amongst involved parties is an extremely solid barrier to enhancing SC systems. Therefore, most participants in the system mainly depend on mediators (also known as third parties) to act as trust agents and to check business dealings, which radically upsurge operating costs and decrease the performance of SC processes.

- Old-fashioned data communication: In modern SCs, information is exchanged among several stakeholders through paper-based records. Generally, vital records, including bills of lading, insurance policies, invoices, credit letters, and different certificates, must transfer with goods. For example, about 200 interactions were essential for Maersk, an international logistics and transport enterprise, to achieve a single consignment of frozen goods to Europe from Mombasa in 2014 (IBM, 2017). These interactions generated a pile of records around 25 cm in height. Due to this obsolete and unproductive data communication, aircraft and ships are frequently deferred in ports if the documents miss-match the carried products.

- Agreement policies: At present, companies have to follow ever more stringent regulatory policies to enable better-quality services and safe products to consumers. In recent times, the U.S. Food and Drug Administration and Federal Trade Commission espoused many policies to improve food security and provide complete visibility of foodstuff flows in the SC. But, using existing SC operations, it is challenging to get these statistics from different involved parties and to create a database that conforms to new policies.

The COVID-19 pandemic has interrupted global SC systems in an unprecedented manner. Poor supply-demand trade-offs have pushed global industries to struggle with numerous challenges like increasing product costs, port bottlenecks, logistics disruption, and fluctuating consumer behavior. The pandemic imposes the following potential challenges to sustainable supply networks (Okorie et al., 2020):

- Material shortage: Shortage of raw materials and accessories in the market owing to outpourings in demand as well as panic purchasing combined with radical uncertainty on the suppliers' side, such as the absence of imported products due to nationwide lockdowns.
- Supply unpredictability: Uncertainty and constraints at the dealer's side combined with unpredictability in cost and amount of critical raw material cause certain obstacles to the SC management system.
- Uncertainty in demand: Unpredictability and irregularity in orders obtained from consumers owing to changes in purchasing activities, including a decrease in procuring of luxury and precious products, absence of cognizance about coronavirus, and non-consumption of some food products initiating demand interruption.
- Delayed delivery: Limitations on imports and local conveyance associated with particular directions containing detours owing to containment regions, slower transportation of products, more lead times, consequently affecting the timely distribution of vital commodities as well as raw materials.

- Shortage of workforce: Forced shutdowns cause a decrease in remunerations, dearth of sustenance and employment issues, leading to trained employers to migrate back to their corresponding places, causing a shortage of labor and a key obstacle to effective SC processes in a developing economy like India.

- Substandard manufacturing practices: Most of the production hubs across the globe are situated in areas that were considered containment zones. Hence, the complete shutting of these hubs creates a massive challenge to manufacturing activities. Moreover, variable demand and unpredictable supply caused the production of a substandard product portfolio bringing more complications to the SC.

- Limitations in storage capacity: Decreasing customer or business trust, panic buying, deficiency of demand for luxury products causing the overstocking of goods in distribution hubs and local storerooms. Furthermore, the users are anticipated to decrease their outlay for a long time. This leads to blockage of funds and cash flow problems. Considering the above-mentioned issues including, overstocking, inventory pile-up, and blockage of funds are the main barrier to effective SC processes.

- Vulnerable transportation system: Absence of viable automobiles traveling between major routes, unadorned constraints on exports, and indigenous transportation along with distribution ways in containment areas causing transportation inaccessibility and deferrals in delivery.

- Fostering of substandard alternatives: Due to the lockdown restrictions for critical suppliers and the ease of access of substandard substitutes/alternatives in the market some companies get such alternatives. This leads to poor quality, additional costs, rework, and long-term adverse impacts on SC.

- Last-mile distribution problems: Most metropolitan cities, which comprise the mainstream of the inhabitants, are considered containment areas. Changing routes to avoid these areas causes in-transit deferrals. Indigenous and state-level guidelines and deferments related to e-pass issuance, validity, and compliance contribute to last-mile distribution challenges.

In the aftermath of severe disruption from the COVID-19 contagion, global organizations plan to shake up their SC policies to become more cooperative, resilient, and interconnected with consumers, vendors, and other involved parties. To do that, they will upturn investment in innovative SC systems including Blockchain technology. The following section analyzes the driving and restraining forces to fostering BCT in a sustainable supply network in the context of the pandemic outbreak.

5.4 Driving Forces of BCT Fostering in Sustainable SC Management

BCT can have revolutionary and disruptive implications for SC practices. However, planning to foster the BCT is indispensable in several industries, predominantly in sustainable SCs management systems. BCT leverages several potential reimbursements and plays a key role in tracing, controlling, and mainly securing commodities in the supply network. In this work, we study the driving and restraining forces of BCT-based technology adoption in sustainable SC networks. The restraining forces decelerate the revolution, neutralize the motivating forces and competencies of BCT, and hamper effective transformation within sustainable SCs and organizations. The information systems and digital technologies of SC endure to act as vital roles in the SC management system. Conventional schemes might not handle numerous multifaceted and active problems associated with modern SCs. Most of these networks fail to offer rationalized, secured, real-time information. BCT comprises abundant proficiencies to aid current sustainable SCs. Improved security and trust, complete transparency and auditability, and disintermediation are some standard enablers for BCT fostering (Saberi et al., 2019).

The robust reason that the BCT is the key instrument of the supply network is its exclusive characteristics such as reliability, transparency, traceability, real-time data exchange, visibility, and cyber security, all of which increase the effectiveness of the supply network. These characteristics can increase the scope of the SC management, operational productivity and can transform from conventional to rationalized SCs (Morkunas et al., 2019). For instance, in multifaceted SC management systems, various suppliers contribute to the SC procedures, and a common database is essential to preserve all transactions. Consequently, an efficient digital communication model is vital that exchanges of secure and real-time data. BCT can increase the integration and cooperation among SC operations. Table 5.1 shows the significance of BCT characteristics including transparency, auditability, versatile value sharing, improved data security, reliability, and visibility in SC management processes and organizational performance.

5.5 Restraining Forces of BCT Fostering in Sustainable SC Management

At present, the popular research literature studies are brimming with the benefits of BCT applications and are frequently used for sustainable SC. As mentioned in previous sections, BCT can increase the performance of the

TABLE 5.1

Significance of BCT Features in Organizational Performance and SC Management Practices

Features of BCT	Significance in Organizational Performance	Significance in SC Practices	References
Transparency	Transparency of BCT improves the effectiveness of the SC processes by enhancing inventory planning as well as minimizing the transaction time using automation, timestamping, and the real-time auditing system.	This feature enables the transparency of the transaction and product by creating a strong correlation between involved parties such as consumers, suppliers, procurement, and subcontractors.	Kim and Shin (2019), Cole et al. (2019)
Auditability	This feature enables complete traceability for inventory and efficient resource allocation policies which enable a cost-effective, predictable, flexible, and resilient manufacturing system.	BCT offers a better traceability model and increases the auditability and transparency of raw materials, accessories, goods, and data among SC processes.	Hasan et al. (2020)
Versatile value sharing	This feature aids to improve planning and flexibility and decreased lead time. Also, it is used for predicting, utilizing, and allocating organizational resources, services, and competencies effectively.	BCT improves strategic planning, SC integration, and creates an efficient and better correlation between SC processes and stakeholders.	Kshetri (2018), Hald and Kinra (2019)
Improved data security	This feature enhances the intra- and inter-organizational trust level and privacy with a lucrative method with secured data and increased immutability.	Improved information security ensures a protected transactions environment and improved trust level among SC participants.	Hald and Kinra (2019), Cole et al. (2019), and Kim and Shin (2019)
Reliability	This feature guarantees the security and dependability of information which is useful to increase the performance and trust level of the firm.	BCT increases the dependability of the SC operations and processes such as strong correlations with SC participants.	Kshetri (2018), Hasan et al. (2020)
Visibility	This feature increases collaboration between organizations and stakeholders and improves the visibility of organizational practices and records.	BCT enables the immutable identity of supplies for enhancing the SC processes with dependable methods.	Kim and Shin (2019), Kshetri (2018)

SC system. However, any novel technology fostering is replete with hitches; BCT is not absolved. Any innovative tool can harvest the benefits only when its potential restraining forces are overcome. The stakeholders of SC should intensely know these difficulties and design accordingly. In BCT, distributed records demand greater processing capacity and resources for preserving the security of records and data that are reproduced, which eventually cause higher energy consumption (Zhou et al., 2020). Higher processing capacity is mandatory for significant PoW consensus mechanisms which ingest considerable energy. This higher power dissipation also leads to more carbon discharges.

In this work, we identify various restraining forces and challenges for BCT fostering in the sustainable SC management context. These restraining forces can be classified into three major types based on environmental, organizational, and technological contexts. The environmental context refers to the barriers related to features of industries, markets, and the regulatory environment. The organizational context refers to forces due to the structure of the organization, the resources, and intra-organizational interactions. The technological context includes the forces due to features and availability of technological innovation (Baker, 2012).

5.5.1 Restraining Forces from Environmental Obstacles

The environmental contexts comprise the restraining forces from the market competitiveness, industry features, regulatory environment, and the correlation among organizations. We used the restraining forces pinpointed by Saberi et al. (2019) and further extended and assimilated them with the state-of-the-art literature. The following factors are considered as the major restraining forces in fostering BCT in SCs:

- Lack of understanding: Lack of customer awareness about fostering BCT for sustainable SC processes owing to unsuccessful interaction and cooperation among the participants with different preferences and objectives. Companies often lack sustainability awareness aid and fail to foster viable processes across their SC; in this case, the BCT only increases the overheads and potential confusion (Mangla et al., 2017).
- Challenge of data privacy: SC stakeholders might have diverse privacy requirements and various strategies related to data and information employed in sustainable SCs and for BCT. Privacy, confidentiality, and the importance of information may be concerns (Wang et al., 2019).
- Challenges in technology integration: Integrating sustainability processes and BCT with traditional SC practices is a perplexing endeavor. Furthermore, processes, materials, and technology

evolution are required to aid sustainable methods (Mangla et al., 2017). For example, technologies and services need to be upgraded to connect Internet-of-things for collecting data from them (Morkunas et al., 2019).

- Synchronization problems: Lack of cooperation, interaction, and synchronization among SC stakeholders with diverse and often conflicting incentives/goals and preferences (Kamble et al., 2019; Kshetri, 2018).

- Cultural diversity: Various topographical or structural cultures of SC stakeholders can impede BCT adoption in sustainable SC practices (Wang et al., 2019).

- The dearth of organizational policies: Some governments might be unwilling to foster BCT and sustainable SC procedures (Biswas & Gupta, 2019; Kamble et al., 2019)

- Competition intensity and uncertainty: Implementing sustainable SC processes and BCT is onerous. It may disturb the market attractiveness and bring related threats to the firms. For example, unpredictable fluctuations in customer demands, the behavior of the consumer, and imminent transactions (Wang et al., 2019; Biswas & Gupta, 2019).

- The dearth of contribution: Deficiency of contribution and contradictory goals of associated communities and stakeholders to aid sustainable processes and BCT (Wang et al., 2019).

- Shortage of industry participation: Deficiency of industry guidance in safe and ethical procedures for implementing BCT in sustainable SCs (Hughes et al., 2019).

- Deficiency of incentives and rewards: Difficulties in motivating sustainable processes and BCT; or absence of reward systems to guarantee the data integrity and incentivize these processes by professional and government sectors (Wang et al., 2019).

5.5.2 Restraining Forces from Organizational Obstacles

The organizational context includes restraining forces and challenges related to organizational obstacles. The following factors are considered as the major restraining forces in fostering BCT in SCs in terms of organizational context:

- Economic limitations: BCT needs software and hardware, with maintenance, to sustain it. The budget related to added investments rises with the bigger application (Saberi et al., 2019). Innovative technologies will be overpriced for the firm and the participants, not only for the technology but for supporting stakeholders and operational substructure (Morkunas et al., 2019). Besides, data ingestion through SC and constructing new business models enact outlays on companies

(Biswas & Gupta, 2019). Moreover, fostering sustainable processes is expensive. Corporates are restricted in economic resources to foster BCT in a sustainable SC management system.

- Deficiency of management responsibility and sustenance: Many administrators fail to possess long-haul support and responsibility for sustainable SC procedures and espousing unruly technology (Wang et al., 2019). The deficiency of responsibility from middle or top-level management imposes challenges. Their support is indispensable for BCT solicitations in sustainable SCs (Mangla et al., 2017).

- The dearth of new corporate policies: Companies need to describe new strategies to accept BCT (what is the correct convention of the technology, for instance, when and where) (Wang et al., 2019).

- Capability shortage: Deficiency of technical knowledge and expertise about BCT and sustainable SCs is another big issue (Kamble et al., 2019). This discomfort, used to a comparatively new corporate procedure including sustainability, adversely impacts the perceived accessibility of the organizational resources (Kamble et al., 2019).

- Difficulty in shifting corporate culture: Espousing BCT changes or renovating current corporate standards. Corporate standards include strategies for work culture and suitable behavior of the companies. There are difficulties in implementing BCT in SCs owing to a deficiency of regularization (Morkunas et al., 2019). Intra-firm deviations for an innovative culture, both in BCT and in sustainable SC, would add challenges to developing relationships between companies.

- Reluctance to adopt new technology: Fostering new technology would need switching or changing legacy systems. This dispute may lead to hesitation and resistance from corporates and businesses (Saberi et al., 2019).

- Lack of tools: Absence of methods and apposite approaches, techniques, measures, and tools for BCT application and calculate sustainability enactment within organizations (Morkunas et al., 2019).

5.5.3 Restaining Forces from Technological Obstacles

The restraining forces related to technical context contain technological al competence, complication, effort, limitations, and accessibility of the Blockchain that is considered for fostering in SCs. We summarize the restraining forces from these obstacles are given below:

- Immaturity of technology: Indeed, BCT is immature. This immaturity generates serious issues such as usability, scalability, and compatibility problems (Biswas & Gupta, 2019). Moreover, The BCT cannot handle large numbers of transactions effectively. Similarly, storing large blocks is a difficult endeavor, managing real-time big data (e.g., "bloat" issue in Bitcoin) (Biswas & Gupta, 2019).

- Latency and throughput issues: BCT still faces throughput and latency problems (Swan, 2015). With a minimum throughput rate and greater delay, BCT still needs improvement. These concerns show that deployment of BCT in sustainable SC could mean less number of transactions, and the transaction times might be more. When planning to oversee social and environmental processes, the category, location, and amount of data make it really hard to handle.

- Security breaches: BCT has been presented as a secure technology that uses an exclusively distributed configuration with different data processing algorithms that make it hard to crash or hack. However, innumerable cyber-attacks, access to sensitive data, false data injection activities, and other system threats, particularly in transactions with cryptocurrency, have raised queries about the susceptibility of BCT adoption in sustainable SC systems (Wang et al., 2019).

- Access to technology: Information and communication technology (ICT) infrastructure and the Internet are significant resources for BCT fostering. In many instances, the ICT structure of a firm is deprived or technology accessibility is impossible.

- Immutability challenge: Immutability denotes that information in records cannot be removed or modified. It is a powerful feature of BCT that guarantees the dependability and validity of data. However, if inappropriate data fed into the BCT can be modified with supplementary data, the record history of the transaction will always be in the BCT (Biswas & Gupta, 2019). For instance, poor social or environmental information could be present incessantly, although the up-to-date data pursues to correct such data.

- The adverse portrayal of technology: This feature is not firmly technological; however, perception acts a vital role in ultimate implementation. The public may associate BCT primarily with cryptocurrencies like Bitcoin or crypto-assets. These developments might be observed as malevolent actions in some countries. Consequently, companies may hesitate to foster general BCT in their sustainable SC processes (Swan, 2015).

5.6 Conclusion

The pandemic has exposed numerous cracks and weaknesses in the conventional pattern of manufacturing, consumption, and their ongoing effect on supply networks. During the pandemic, the potential of digital technology initiatives in the current state of sustainable SC systems became apparent. Several organizations are fostering BCT to solve the pandemic distribution and set forth to handle the sustainable processes of their supply networks. The key characteristics of BCT such as greater traceability, improved transparency, increased efficiency, secured certification, and proof of origin for goods on top of billions of dollars in commercial savings make BCT an indispensable part of current sustainable SC to resolve the problems including market monopolies, data integrity, and ownership. In spite of the potential benefits of BCT, the speed of fostering in sustainable SC management system has been decelerated. These applications are halted at the trial phase. In this chapter, we explore how fostering BCT in sustainable SC processes with several environmental, organizational, and technological promises have been delayed. Also, this chapter will carry out a systematic study of BCT technology, sustainable SC, and driving as well as restraining forces for espousing BCT to achieve SC with improved sustainability. The major limitation of this research is that the established correlation between BCT characteristics and SC processes is based on literature only and could not be described experimentally. Hence, we plan to implement BCT in real-time in order to study the performance of the organization by adopting BCT into SC processes,.

References

Abu-elezz, I., Hassan, A., Nazeemudeen, A., Househ, M., and Abd-alrazaq, A. 2020. The benefits and threats of blockchain technology in healthcare: A scoping review. *International Journal of Medical Informatics*, 142, 104246.

Alladi, T., Chamola, V., Parizi, R. M., and Choo, K. K. R. 2019. Blockchain applications for industry 4.0 and industrial IoT: A review. *IEEE Access*, 7, 176 935–176 951.

Baker, J. 2012. The technology–organization–environment framework. In: *Information Systems Theory*. Springer, pp. 231–245.

Baliga, A. 2017. *Understanding Blockchain Consensus Models*. Whitepaper, pp. 1–14.

Biswas, B., and Gupta, R. 2019. Analysis of barriers to implement blockchain in industry and service sectors. *Computers and Industrial Engineering*, 136, 225–241.

Capgemini Research Institute. 2020. *Fast Forward – Rethinking Supply Chain Resilience for a Post-COVID-19 World*. https://www.capgemini.com/wp-content/uploads/2020/11/Fast-forward_Report.pdf

Casey, M., and Wong, P. 2017. Global supply chains are about to get better, thanks to blockchain. *Harvard Bussiness Review*, 13, 1.

Clavin, J., Duan, S., Zhang, H., Janeja, V. P., Joshi, K. P., and Yesha, Y. 2020. Blockchains for Government: Use cases and challenges. *Digital Government: Research and Practice*, 1(3), 1–21.

Cole, R., Stevenson, M., and Aitken, J. 2019. Blockchain technology: Implications for operations and supply chain management. *Supply Chain Management*, 24(4), 469–483.

Dutta, P., Choi, T. M., Somani, S., and Butala, R. 2020. Blockchain technology in supply chain operations: Applications, challenges and research opportunities. *Transportation Research: Part E, Logistics and Transportation Review*, 142(2020) , 102067.

Fernandez-Carames, T. M., and Fraga-Lamas, P., 2018. A review on the use of blockchain for the internet of things. *IEEE Access*, 6, 329–330.

Gartner.2019. *Gartner Predicts 90% of Current Enterprise Blockchain Platform Implementations Will Require Replacement by 2021.* https://www.gartner.com/en/newsroom/press-releases/2019-07-03-gartner-predicts-90--of-current-enterprise-blockchain

Gupta, M. 2018. *Blockchain for Dummies.* 2nd IBM Limited ed. Hoboken, NJ, US: John Wiley & Sons, Inc.

Hald, K. S., and Kinra, A. 2019. How the blockchain enables and constrains supplychain performance. *International Journal of Physical Distribution & Logistics Management*, 49(4), 376–397.

Hasan, M. R., Shiming, D., Islam, M. A., and Hossain, M. Z. 2020. Operational efficiency effects of blockchain technology implementation in firms: Evidence from China. *Review of International Business and Strategy*, 30(2), 163–181.

Huang, C., Wang, Y., Li, X., Ren, L., and Zhao, J. 2020. Clinical features of patients infected with 2019 novel coronavirus in Wuhan, China. *The Lancet*, 395(10223), 497–506.

Huang, R. 2019. UN pilot In: *Mongolia Uses Blockchain to Help Farmers Deliver Sustainable Cashmere.* Available Online: https://www.forbes.com/sites/rogerhuang/2019/12/28/un-pilot-in-mongolia-uses-blockchain-to-help-farmers-deliver-sustainable-cashmere/?sh=16a2516617d9

Hughes, L., Dwivedi, Y. K., Misra, S. K., Rana, N. P., Raghavan, V., and Akella, V. 2019. Blockchain research, practice and policy: Applications, benefits, limitations, emerging research themes and research agenda. *International Journal of Information Management*, 49, 114–129.

IBM.2017. *I.N. Release. Maersk and IBM Unveil First Industry Wide Cross Border Supply Chain Solution on Blockchain.* Available Online: https://www.03.ibm.com/press/us/en/pressrelease/51712.wss#feeds

Ivanov, D., and Das, A. 2020. Coronavirus (COVID-19/SARS-CoV-2) and supply chain resilience: A research note. *International Journal of Integrated Supply Management*, 13(1), 90–102.

Kamble, S., Gunasekaran, A., and Arha, H. 2019. Understanding the blockchain technology adoption in supply chains—Indian context. *International Journal of Production Research*, 57(7), 2009–2033.

Kim, J. S., and Shin, N. (2019). The impact of blockchain technology application on supply chain partnership and performance. *Sustainability*, 11(21).

Kosba, A., Miller, A., Shi, E., Wen, Z., and Papamanthou, C. 2016. Hawk: The blockchain model of cryptography and privacy-preserving smart contracts. In: *2016 IEEE Symposium on Security and Privacy (SP)*, pp. 839–858.

Kshetri, N. 2018. Blockchain's roles in meeting key supply chain management objectives. *International Journal of Information Management*, 39, 80–89.

Litke, A., Anagnostopoulos, D., and Varvarigou, T. 2019. Blockchains for supply chain management: Architectural elements and challenges towards a global scale deployment. *Logistics*, 3(1), 5.

Mangla, S.K., Govindan, K., and Luthra, S. 2017. Prioritizing the barriers to achieve sustainable consumption and production trends in supply chains using fuzzy analytical hierarchy process. *Journal of Cleaner Production*, 151, 509–525.

Maouchi, M., Ersoy, O., and Erkin, Z. 2018. TRADE: A transparent, decentralized traceability system for the supply chain. In: *Proceedings of 1st ERCIM Blockchain Workshop 2018. European Society for Socially Embedded Technologies (EUSSET)*.

Mitselmakher, K. 2018. *The Future of Blockchain Technology: Top Five Predictions for 2030.* https://www.blockchain-expo.com/2018/10/blockchain/future-of-blockchain-technology/# (accessed on 6 February 2022).

Morkunas, V. J., Paschen, J., and Boon, E. (2019). How blockchain technologies impact your business model. *Business Horizons*, 62(3), 295–306.

Murthy, C. V. N. U. B., Shri, M. L., Kadry, S., and Lim, S. 2020. Blockchain based cloud computing: Architecture and research challenges. *IEEE Access*, 8, 205190–205205.

Nakamoto.2008. *Bitcoin: A Peer-to-Peer Eelectronic Cash System.* https://bitcoin.org/bitcoin.pdf

Narayan, R., and Tidström, A. 2020. Tokenizing coopetition in a blockchain for a transition to circular economy. *Journal of Cleaner Production*, 263, 121437.

Okorie, O., Subramoniam, R., Charnley, F., Patsavellas, J., Widdifield, D., and Salonitis, K. 2020. Manufacturing in the time of COVID-19: An assessment of barriers and enablers. *IEEE Engineering Management Review*, 48(3), 167–175.

Pal, A., Tiwari, C. K., and Behl, A. 2021. Blockchain technology in financial services: A comprehensive review of the literature. *Journal of Global Operations and Strategic Sourcing*, 14(1), 61–80.

Pilkington, M. 2016. Blockchain technology: Principles and applications. In: *Research Handbook on Digital Transformations*. Elgaronline, p. 225.

Popper, N., and Lohr, S. 2017. *Blockchain: A Better Way to Track Pork Chops, Bonds, Bad Peanut Butter*. New York Times, p. 4.

Rejeb, A., Keogh, J. G., and Treiblmaier, H. 2019. Leveraging the Internet of Things and blockchain technology in supply chain management. *Future Internet*, 11(7), 161. doi:10.3390/fi11070161

Saberi, S., Kouhizadeh, M., Sarkis, J., and Shen, L. 2019. Blockchain technology and its relationships to sustainable supply chain management. *International Journal of Production Research*, 57(7), 2117–2135.

Seuring, S., Sarkis, J., Müller, M., and Rao, P. 2008. Sustainability and supply chain management – An introduction to the special issue. *Journal of Cleaner Production*, 16(15), 1545–1551.

Solleder, O., and Velasquez, M. T. 2020. *Blog: The Great Shutdown: How COVID-19 Disrupts Supply Chains, International Trade Centre.* Available Online: https://www.intracen.org/covid19/Blog/The-Great-Shutdown-How-COVID19-disrupts-supply-chains/ (accessed on 22 February 2022)

Swan, M. 2015. *Blockchain: Blueprint for a New Economy*. Sebastopol, CA: O'Reilly Media, Inc.

Uddin, M. A., Stranieri, A., Gondal, I., and Balasubramanian, V. 2021. A survey on the adoption of blockchain in IoT: Challenges and solutions. *Blockchain: Research and Applications*, 2(2), 100006.

Wang, Y., Singgih, M., Wang, J., and Rit, M. 2019. Making sense of blockchain technology: How will it transform supply chains? *International Journal of Production Economics*, 211, 221–236.

Zhou, Q. H., Huang, H. W., Zheng, Z. B., and Bian, A. J. 2020. Solutions to scalability of blockchain: A survey. *IEEE Access*, 8, 16440–16455.

Section 3

Opportunities for Future

6

Opportunities and Challenges for Society 5.0 in Procurement

Atour Taghipour, XiaoWen Lu, and Mustafa Smahi

CONTENTS

6.1 Introduction

Still, supply chain management, especially procurement is evolving. So, it's essential to answer to the new changes and discover the possibilities for company's elaboration. We can observe various digital transformations, however as Parida et al. (2019) showed an agreed description of digitalization is presently missing. The authors explain digitalization as "the use of digital technologies to introduce a business model and give new profit aqueducts and value-producing openings in artificial ecosystems". We find the below description and, in this relation, digitalization concern the procurement is also considered.

An intriguing point occurs in the different stages of e-procurement. The digital procurement can be enabled using blockchain, data analytics, optimization, and collaborative robots. The focus of procurement will be on collaboration to build innovative models (Taghipour, 2009).

Procurement is a part of the value chain that can be changed by digitalization. Srai et al. (2019) defined two types of digitalization technologies used by

DOI: 10.1201/9781003177432-9

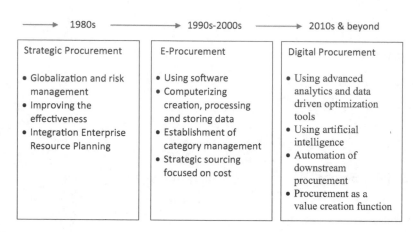

FIGURE 6.1
Procurement evolution. (Source: Kearney analysis.)

supply chains. The first type uses generally the Internet, the other type uses advanced technologies (e.g., blockchain; Figure 6.1).

6.2 The Relation Procurement/Digitalization

Procurement 5.0 contributes to the performance and improves the productivity of the supply chains as efficient as possible. Due to digitalization, procurement 5.0 connects the actors of the supply chains and enables the collaboration in the procurement process. Digitalization contributes to the value-adding process in the supply chains. Robotization can help to decrease purchasing time and optimize the resources. There are many ways to use digital procurement (Merimi and Taghipour, 2021; Taghipour, 2021).

The digitalization of procurement continues to be in its immaturity in some companies (Figure 6.2). It can bring benefits in decision-making processes. In future, procurement becomes a crucial element in strategic decisions to guarantee effectiveness. Digitalization offers numerous openings for companies. On the opposite hand, not keeping step with the changing environment will result in deficiency (Taghipour and Frayret, 2010).

Digitalization must consider the supply chain as an entire part. Within the area of procurement, this means considering all suppliers in a supply chain. In this case, an optimal solution is attainable, for all actors of the supply chain.

The full process is veritably demanding on the degree of information (Ren et al., 2016). It is necessary to decide on an acceptable platform. The platform can assure the speed of calculation, investment cost, deployment, etc.

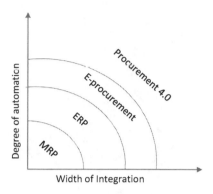

FIGURE 6.2
The evolution of procurement. (Glas and Kleeman, 2016.)

It is important to work out the compass of digitalization and select a reliable provider.

A new definition of digitalization is 'the way numerous disciplines of social life are restructured around digital communication and media architectures', in discrepancy to the affiliated term digitalization meaning the material process of converting analog aqueducts of information into digital bits' (Brennen and Kreiss, 2016). Siu and Wong (2016) explained that 'digitalization refers to the technology of digitalizing information'. Simply, the digitalization implies the use of digital technologies in the service of the businesses. According to Brennen and Kreiss (2016) and Hagberg et al. (2016), digitalization/digitalization 'refers astronomically to the blending of digital technologies into merchandising'.

The use of digitalization requires the understanding of the digital transformation (Figure 6.3). The frame proposed by Kane et al. (2016) provides a constructive information in relation to the use of the technologies. In another work, Hofmann and Rüsch (2017) suggested a table of digital technologies and related literature. In this case, Assiduity 4.0 appears to be a recognized concept within the literature (e.g., Hermann et al., 2016; Hofmann and Rüsch, 2017). In fact, the literature on Assiduity 4.0 talks about different technologies enabled by the Internet, such as pall and mobile computing, mortal commerce, big data, security, social networks, 3D printing, intelligent agents, collaborative robots, etc. (Hozdić, 2015; Wan et al., 2015; Wang et al., 2016; Lu, 2017).

Some authors highlight the arising blockchain technology in relation to Assiduity 4.0 (Hofmann and Rüsch, 2017). We can conclude that the use of digitalization depends on technological foundations. For example, De Boer et al. (2002) identified MRO (via the Internet), e-sourcing (using Internet technology), e-tendering (via the Internet), e-reverse auctioning (via the Internet). Johnson and Klassen (2005) differentiated between sourcing and

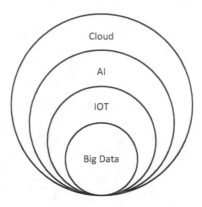

FIGURE 6.3
Design model for procurement 4.0. (Fröhlich et al., 2018.)

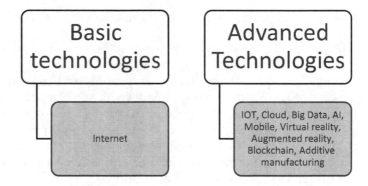

FIGURE 6.4
The PSM (purchasing and supply management) digitalization grid, populated with extant research. (Inspired from Srai and Lorentz, 2019.)

coordination (automated processes). Presutti (2003) distinguished between e-design, e-sourcing, and e-transactions (the robotization; Figure 6.4).

The actors of supply chain produce an important amount of data (Allal-Chérif et al., 2021). These data play an important role in performance of the supply chains. But, the data produced aren't exploited perfectly (Mikalef et al., 2018). Unwillingness to change is another problem in digital transformation. To be competitive the companies need to do further than just survive. In this case, the procurement department plays an important role. Digitalization can give supply chains with the means to forecast future trends and to manage better the operations (Borges et al., 2009). Digitalization can use a variety of technics, from quantitative to qualitative approaches. As an example, robotization optimizes the processes to obtain the best results (Allal-Chérif and Maira, 2011). AI can help the decision makers to analyse an important amount of data at a lower cost (Salminen et al., 2017). The procurement function can analyse multidimensional data to take an optimal decision (Baryannis et al., 2019). Therefore, access to the new technology is essential, because without using intelligent systems it's nearly impossible.

To switch to the digital procurement, the company needs to achieve the maturity step. A second step is the state of independency, as a separate entrepreneurial function. At this step, the focus is on cost-effectiveness. The integration between different functions is crucial, and supply chain operation processes are defined and proved (Lebosse et al., 2017). The third step is *Supporting*, which means supporting of the company's competitive strategy. The fourth step is the total Integration, which means integration with suppliers is based on a long-term relationship. It is essential to measure the performance at each step.

The original abstract model of Procurement 4.0 is grounded on four fundamentals motorists of digital metamorphosis. The use of new technologies is changing very fast. In the future, an important improvement in the supply chains is anticipated. McKinsey studies show that digitalization can offer a reduction to 40 of the total cost of procurement. In addition, the time-to-request will be reduced to 50% and the trust ability increases to 85%.

The fast advances in digitalization are modifying supply chains. There exist different digital tools and knowledge technologies, which help the procurement. Stephens et al. (2013) accentuated the growing significance of digitization within the procurement, which can impact the entire supply chain. The companies can use several varieties of procurement technologies such as e-sourcing, e-contracts, e-catalogs, e-invoicing, etc. These new arising technologies are embodied in an exceedingly simple term called "Industry 4.0". This generality has been initiated to improve the coordination of the actors of supply chain. To realize this, several technologies are being adapted to enable the procurement function.

6.3 New Technologies in Procurement

These technological improvements are fundamental tools that can impact the procurement function to achieve the global performance for the entire supply chain.

6.3.1 Big Data

Enterprises produce an important volume of knowledge, which is usually stored in data in each function. The use of big data helps companies to improve their capabilities and speed up the process of decision-making. Big data is an enabler technology, which structuring, simplifying, and analysing the information. Through the applicable operation of those technologies, both enterprises (buyers) and merchandisers (suppliers) establish a 'palm–palm' relationship. This could help in delivering advancements within the network's effectiveness, effectiveness, and dexterity.

6.3.2 Robotization

Presently, robotics is one of the most important enablers in mechanical systems such as automobile industry. These technologies have the potential to accelerate procurement decisions by automatizing transactional decisions (Cauchois et al., 2017). The utilization of automated machines inside the factories can help the procurement function by increasing the visibility across the supply chains (Figure 6.5).

6.3.3 The Internet of Things (IoT)

The Internet of Things is a fundamental enabler to connect physical objects. It uses sensors, detectors, and communication devices into physical objects (e.g., wireless detector, GPS, etc.) that help to connect physical objects. The connected objects can be tracked and coordinated and controlled in a digital manner. This technology uses collected data through Internet from individual sources. So, information is necessary.

This technology can help the procurement function by decreasing the complexity of supply chains. As a result, this allows the procurement function to communicate in real time. Also, GPS devices and connected sensors give access to the placement of the cargo and carrying condition. The coordination can be streamlined and conflicts can be reduced (Figure 6.6).

6.3.4 Blockchain

A blockchain is a decentralized system that can establish a trusted environment for exchanging information across multitudinous actors (i.e., bumps). It

FIGURE 6.5
Supporting procurement function with robotics. (Taghipour and Frayret, 2013.)

FIGURE 6.6
Big data analytics. (Ahmed et al., 2019.)

is viewed as a group of tools that address a specific business problem. One of the most important roles of blockchain in purchasing is to help all actors of supply chain involve within decision-making processes. In this centralized platform, companies can save the data related to the suppliers, the agreements, and etc. So, using blockchain technology can be the source of optimization in supply chains (Shin and Taghipour, 2021).

A blockchain can contribute in the process of selecting the suppliers to the payments (Mbiatem et al., 2018). Procurement directors will communicate the backlog to achieve an insight into the whole supply chains. Blockchains transform rapidly the process of digitalization into a competitive advantage for enterprises. In traditional force chains, companies generally capture data on batches of product employing a combination of systems and manual paper trails. That is, paper-predicated system that passes the data together with the products or goods. This can be constantly tractable with the nonsupervisory demand for 1-up/1-down traceability within the food and consumer good sectors as an illustration. In doing so, whenever goods and combined documentation (e.g., bills of lading or boat advertisement) pass from one actor within the force chain to a different one, these documents are in a pitfall of counterfeiting, impurity, or theft. In addressing these issues, a configured blockchain helps to grease and secure the exchange of sourcing documents, guaranties, product assurances (i.e., organic, kosher, halal, etc.), and related specifications as inflexible, timestamped, certified, and accessible to the interested realities. Still, all data inputs into a blockchain record must be vindicated to palliate the pitfall of 'garbage-in, garbage-out' (Figure 6.7).

6.3.5 Smart Contracts

Some operations of blockchain technology introduce the generality of "smart contracts". N. Hackius and M. Petersen (2017) preceded bitcoin's operation by more than a decade. Smart contracts are part of a blockchain, first introduced by Nick Szabo in 1994. Szabo defined it as "a motorized trade protocol that executes the terms of a contract". Likewise, he suggests that the contracts should be integrated into software law and automated programs within a blockchain, which can be executed when certain conditions are met. They are simple scripts located on the blockchains to perform certain operations.

From the purchasing point of view, smart contracts concern to any purchasing deals. They can automate business processes. The smart contracts can ensure the payments, if the conditions are respected.

FIGURE 6.7
Leveraging procurement process with blockchain technology. (Nicoletti, 2017.)

6.4 Conclusion

Procurement is the main part of any supply chain. In this context, the new technology can help the procurement function to play an important role in the competitiveness of supply chains. The main question is which technology can be adapted to the procurement in order to improve the efficiency of the supply chains (Loivet et al., 2020). The effectiveness of new technology lies in the simplification of processes, process automation and centralization of all purchasing, which is the key success in the optimization. We conclude that adapting new technologies can help the procurement to move from a traditional focus to a strategic focus in supply chains.

Digital procurement is considered as revolution supply chains. However future examination is required to estimate the cost, benefit, and volume analysis, as well as a feasibility analysis in different types of companies.

References

Ahmed, M., Choudhury, S., & Al-Turjman, F. (2019). Big data analytics for intelligent Internet of Things. In *Artificial Intelligence in IoT* (pp. 107–127). Cham: Springer.

Allal-Chérif, O., & Maira, S. (2011). Collaboration as an anti-crisis solution: The role of the procurement function. *International Journal of Physical Distribution & Logistics Management*, 41(9), 860–877.

Allal-Chérif, O., Simón-Moya, V., & Ballester, A. C. C. (2021). Intelligent purchasing: How artificial intelligence can redefine the purchasing function. *Journal of Business Research*, 124, 69–76.

Baryannis, G., Validi, S., Dani, S., & Antoniou, G. (2019). Supply chain risk management and artificial intelligence: State of the art and future research directions. *International Journal of Production Research*, 57(7), 2179–2202.

Borges, M., Hoppen, N., & Luce, F. B. (2009). Information technology impact on market orientation in e-business. *Journal of Business Research*, 62(9), 883–890.

Brennen, J. S., & Kreiss, D. (2016). Digitalization. In The *International Encyclopedia* of *Communication Theory* and *Philosophy* (pp. 1–11). WILEY Blackwell.

Cauchois, R., Taghipour, A., Kang, D., Zoghlami, N., & Abed, M. (2017). Advanced solutions for a supply chain with stochastic information. *Journal of Advanced Management Science*, 5(1).

De Boer, L., Harink, J., & Heijboer, G. (2002). A conceptual model for assessing the impact of electronic procurement. *European Journal of Purchasing & Supply Management*, 8(1), 25–33.

Fröhlich, E., & Steinbiß, K. (2018). Supplier relationship management goes digital: First empirical insights. *Universal Journal of Industrial and Business Management*, 8(3), 6373.

Fröhlich, E., Steinbiß, K., & Whelan, M. W. (2018). Supplier management in the age of digital transformation. *Journal of Engineering, Management and Operations*, 1(1), 153–164.

Glas, A. H., & Kleemann, F. C. (2016). The impact of industry 4.0 on procurement and supply management: A conceptual and qualitative analysis. *International Journal of Business and Management Invention*, 5(6), 55–66.

Hackius, N., & Petersen, M. (2017). Blockchain in logistics and supply chain: trick or treat?. In *Digitalization in Supply Chain Management and Logistics: Smart and Digital Solutions for an Industry 4.0 Environment. Proceedings of the Hamburg International Conference of Logistics (HICL)*, Vol. 23 (pp. 3–18). Berlin: epubli GmbH.

Hagberg, J., Sundstrom, M., & Egels-Zandén, N. (2016). The digitalization of retailing: An exploratory framework. *International Journal of Retail & Distribution Management*, 44(7), 694–712.

Hermann, M., Pentek, T., & Otto, B. (2016, January). Design principles for industrie 4.0 scenarios. In *2016 49th Hawaii international conference on system sciences (HICSS)* (pp. 3928–3937). IEEE.

Hofmann, E., & Rüsch, M. (2017). Industry 4.0 and the current status as well as future prospects on logistics. *Computers in Industry*, *89*, 23–34.

Hozdić, E. (2015). Smart factory for industry 4.0: A review. *International Journal of Modern Manufacturing Technologies*, 7(1), 28–35.

Johnson, P. F., & Klassen, R. D. (2005). E-procurement. *MIT Sloan Management Review*, 46(2), 7.

Kane, G. C., Palmer, D., Phillips, A. N., Kiron, D., & Buckley, N. (2016). Aligning the organization for its digital future. *MIT Sloan Management Review*, 58(1), 1–30.

Lebosse, S., Taghipour, A., & Canel-Depitre, B. (2017). Quick response to fluctuations in supply chains: A review. *Journal of Advanced Management Science*, 5(5), 394–400.

Loivet, W., Taghipour, A., & Kang, D. S. (2020). The rise of green supply chain management: Between complexity and necessity. *Journal of Economics, Business and Management*, 8(1), 1–7.

Lu, Y. (2017). Industry 4.0: A survey on technologies, applications and open research issues. *Journal of Industrial Information Integration*, 1, 6–10.

Mbiatem, J., Taghipour, A., & Canel-Depitre, B. (2018, August). Supplier selection approaches for decision makers. In *Proceedings of the 8th International Conference on Information Communication and Management*, London: *Association for Computing Machinery* (pp. 108–112).

Merimi, M., & Taghipour, A. (2021). Accelerating the digitalization of the supply chain: An empirical research about COVID-19 crisis. In *Digitalization of Decentralized Supply Chains During Global Crises* (pp. 1–24). USA: IGI Global.

Mikalef, P., Pappas, I. O., Krogstie, J., & Giannakos, M. (2018). Big data analytics capabilities: A systematic literature review and research agenda. *Information Systems and e-Business Management*, 16(3), 547–578.

Nicoletti, B. (2018). The future: procurement 4.0. In *Agile procurement* (pp. 189–230). Cham: Palgrave Macmillan.

Parida, V., Sjödin, D., & Reim, W. (2019). Reviewing literature on digitalization, business model innovation, and sustainable industry: Past achievements and future promises. *Sustainability*, 11(2), 391.

Presutti, W. D. (2003). Supply management and e-procurement: Creating value added in the supply chain. *Industrial Marketing Management*, 32(3), 219–226.

Ren, Z., Taghipour, A., & Canel-Depitre, B. (2016, October). Information sharing in supply chain under uncertainty. In *2016 6th International Conference on Information Communication and Management (ICICM)* (pp. 67–71). UK: IEEE.

Salminen, V., Ruohomaa, H., & Kantola, J. (2017). Digitalization and big data supporting responsible business co-evolution. In J. Kantola, T. Barath, S. Nazir, & T. Andre (Eds.), *Advances in Human Factors, Business Management, Training, and Education* (pp. 1055–1067). Cham: Springer.

Shin, M., & Taghipour, A. (2021). *Supply Chains Digital Transformation: Automated Underground Logistics Systems*. Underground Construction, 6, 7.

Siu, K. W. M., & Wong, Y. L. (2016). Learning opportunities and outcomes of design research in the digital age. In *Handbook of Research on Learning Outcomes and Opportunities in the Digital Age* (pp. 538–556). USA: IGI Global.

Srai, J., & Lorentz, H. (2019). Developing design principles for the digitalisation of purchasing and supply management. *Journal of Purchasing and Supply Management*, 25(1), 78–98.

Stephens, J., & Valverde, R. (2013). Security of e-procurement transactions in supply chain reengineering. *Journal of Computing and Information Science in Engineering*, 6(3), 1–20.

Taghipour, A. (2009). Evaluation de la collaboration dans une chaîne d'approvisionnement. *Revue française de gestion industrielle*, 28(1), 26–42.

Taghipour, A. (Ed.). (2021). *Digitalization of Decentralized Supply Chains During Global Crises*. USA: IGI Global.

Taghipour, A., & Frayret, J. M. (2010, October). Negotiation-based coordination in supply chain: Model and discussion. In *2010 IEEE International Conference on Systems, Man and Cybernetics* (pp. 1643–1649). Turkey: IEEE.

Taghipour, A., & Frayret, J. M. (2013, May). An algorithm to improve operations planning in decentralized supply chains. In *2013 International Conference on Advanced Logistics and Transport* (pp. 100–103). IEEE.

Wan, J., Cai, H., & Zhou, K. (2015, January). Industrie 4.0: Enabling technologies. In *Intelligent Computing and Internet of Things (ICIT), 2014 International Conference on* (pp. 135140). IEEE, China.

Wang, S., Wan, J., Li, D., & Zhang, C. (2016). Implementing smart factory of industrie 4.0: An outlook. *International Journal of Distributed Sensor Networks*, 12(1), 3159805.

7

Opportunities and Challenges for Blockchain Technology in Supply Chain Management: Reflection on Society 5.0

K Raj Kumar Reddy and P. Kalpana

CONTENTS

7.1 Introduction

World is going through challenging times, pandemic impacted most of the countries heavily. Almost every region is struggling as the climate is uncertain, geopolitical tensions are heating up day by day, and inflation is going higher. In the entire world, no country is self-sufficient; it requires few goods or commodities or essentials or technology and equipment from other countries, so supply chains (SC) plays a major role in ensuring parity between supply and demand. For example, when the pandemic started in China, many enterprises were impacted and those enterprises never knew they were so dependent on Chinese made; clearly, visibility, transparency, and traceability were lacking in SCs.

DOI: 10.1201/9781003177432-10

Post pandemic, digital technologies are emerging in rapid speed, and many enterprises are leveraging technologies in their practices. Technologies such as Internet of Things (IoT), data science, machine learning, reinforcement learning, deep learning, artificial intelligence (AI), and blockchain technology (BCT) are being utilized to transform traditional practices into digitalization. Trends are changing, people are curious about knowing their consumables and goods, which require tracking the life cycle activities of any product, and tracking them alone is not sufficient until they are being shared through a trusted platform. Moreover, adopting any digital technology into traditional practices is not a cakewalk; it requires stringent evaluation and sharp measures to ensure efficient leverage.

On the other hand, the world is moving toward Society 5.0, which is a transition from the information age. It is looking to build synchronization between physical space with cyberspace, with the main motive of establishing the economical balance [1]. Everyone must agree that digital technologies are existing for greater good; the right usage of any technology will yield better results from the system. So, one must understand the consequences of using technology for a specific application. In the case of SC, needs and pain points should be evaluated against technology and its capabilities. Moreover, Society 5.0 is not only aiming at synchronizing the physical world with the digital one but also intended to solve economic imbalances. Though the world is transitioning faster with these buzzwords, humanitarian aspects of SC are less-focused, since government bodies are lacking in technology advancements. This study focuses on highlighting opportunities and challenges for digital technologies in SC, along with the discussion on how Society 5.0 can be perceived to improve weightage to the humanitarian perspective of SC.

Following contents in this chapter discuss BCT in SC and Society 5.0. The rest of the chapter is organized as Section 7.2 focuses on terms used in this study and deep dives into every aspect i.e., Society 5.0, Data science, BCT, and Blockchain oracles. Section 7.3 deals with BCT in SC and potential use cases along with the suitability of applications to BCT. Sections 7.4 and 7.5 focus on opportunities and challenges for digital technologies in SC. Section 7.6 suggests that Society 5.0 can be leveraged to enhance the humanitarian aspect of SC. Section 7.7 highlights the limitations and future scope of the study, and the study ends with conclusion.

7.2 Terms Used in the Study

7.2.1 Society 5.0

The world has witnessed advancements over time in different perspectives; few call them as revolutions and few other terms call them as trends. Let's see what society advancements are. In the rock ages, people used to hunt and gather their food, that is classified as Society 1.0, the rock age setup. After Society 1.0, people realized about agriculture, where they cultivated their food and started consuming; at the same time, bartering method was popular, this era is known as Society 2.0. Later, people realized the importance of industrialization, where people process and add value to their cultivated products, this is largely known as industrial revolution, i.e., Society 3.0. Setting up the manufacturing setup, processing food, and establishing brand weren't enough. It required more, that's where the world delved into information age, where people started collecting data about everything and anything. In current times, data is everywhere and value is being added in one or another way to stakeholders, this is termed as information revolution or Society 4.0. The world never settles; groups or individuals always want more, this is where Japan introduced Society 5.0, i.e., synchronizing cyberspace with the physical space which is intended to aim at economical parity in the society [1] (Figure 7.1).

Information society or Society 4.0 has limitations in terms of visibility, transparency, and traceability. Data is being collected from the assets (from machineries, devices, people, interactions, etc.), but nobody knows how accurate the data is and how much value can be added to the individual or group through the acquired data, also data monetizing frameworks are poorly

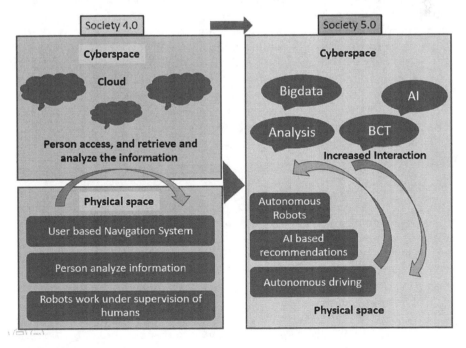

FIGURE 7.1
Society 5.0.

designed, i.e., there is uncertainty around the data value, which eventually impact data quality. Next big shortcoming with Society 4.0 is, that only few are skilled enough to deal with data, so the end user ends up paying more for the service they receive, in fact for too many individuals these digital services are unaffordable. This is where Society 5.0 comes into the limelight, it is a forward-looking society which is looking to automate and synchronize the physical world with cyberspace in every possible way. For example, in current society, the knowledge and information are not being shared across, which leads to locking the value, Society 5.0 frameworks are looking to integrate all practices of data (from data acquisition till consumption) and share them across for value unlocking. Another example will be, though we are living in the information age, many jobs require human intervention, in few cases, humans are being exploited to do repeated tasks, of course human abilities have limitations. To address this Society 5.0 conceptualizes to utilize robots, automated vehicles, etc.

7.2.2 Data Science

The current world is transitioning through the information age, data is being generated from the range of assets, and the same is being utilized to explore & understand the process to a certain extent [2]. Data is the only way to understand the process for the decision-making. Of course, data quality and conformity should be ensured [3]. Many advancements are spanning across the data portfolio, including machine learning, reinforcement learning, deep learning, and ultimately AI [4]. With the availability of the data, predicting the future is very essential, which can foresee the demand, preventive & predictive maintenance, understanding user patterns, etc. [5]. Based on the data availability and complexity of the problem, modeling techniques can be chosen [6]. Though we are living in the information age, every decision-making mechanism is not completely data-driven. There is still a scope to enhance data acquisition and usage.

7.2.3 Blockchain Technology

Blockchain is known for its immutability. Once the data is added to the blockchain, data tampering is near to impossible, as every transaction is mutated in the chain along with timestamps, and a hash of precedent block is utilized [7]. Blockchain is popular because it has no central authority, no participation of trusted third parties or intermediaries, transactions settlements in near real-time, and can produce a high level of transparency [8]. BCT is a type of distributed ledger, the difference comes in the formation of blocks, distributed ledgers do not have chains in block formation. Many correlate blockchain with bitcoin or cryptocurrency, but those are two different things, blockchain is the underlying technology behind cryptocurrencies and many other, countries may restrict usage of cryptos, as they think crypto can be a threat to country's economy, but no country can afford to avoid blockchain, since the advantages of blockchain are very high to ignore [9]. BCT has been around as emerging technology for a while, but it is yet to become as prominent technology. Nowadays, BCT is the backbone for many use cases including decentralized web, decentralized exchanges, decentralized finance (DeFi), Non-Fungible tokens (NFT), stable coins, decentralized identity, blockchain platform as service (PaaS), etc.

With the developments around the world, blockchain is set to become the next big thing from a technology perspective, according to Gartner key growth drivers of blockchain be

1. **Smart contracts** existed for a long time, but it is gaining prominence now. Many are realizing the power of smart contracts and trying to use smart contracts in their potential use cases to add more value to a firm's operations [10]. Typical examples will be blockchain adoption in supply chain management [6] (Food, Automotive, Consumer goods, etc.), identity management, governance-related scenarios, etc.

2. **Adoption of Bitcoin** in financial transactions

3. **Payment network**, distributed ledger technologies are trivial in ensuring fast and safe transactions across the countries, which eventually lead to central banks leveraging blockchain technology to digitize their operations

4. **Decentralized Finance (DeFi)** is an emerging trend, DeFi's are intended to eliminate intermediaries including brokerages, exchanges, banks, etc. from the financial ecosystem through smart contracts running on Blockchain

5. **Tokenization of assets,** the world is running behind NFTs now, many celebrities, creators, and organizations are utilizing NFTs to monetize their assets, still there are uncertainties around taxation and valuation of NFTs. Nowadays, the craze behind NFTs is insane; people are investing heavily without having understanding of the real value of assets, this is pursuing as a set off to volatile crypto market, as purchasing of NFT's are taking place in terms of cryptos. At this moment, the future of the crypto market is uncertain, as many coins are being traded in crypto marketplace, but people are assuming NFT's value will go up significantly in the future, that being the reason people are converting their crypto investments as NFTs.

6. **Blockchain networks** are improving over time (Both private and public blockchain networks), and at the same pace their use cases are also improving. Hyperledger fabric (Private blockchain) is being leveraged for organizational application, Ethereum (Public

blockchain) is accepted widely as programmable money, ether being the cryptocurrency.

7. **Interoperability** is still an issue, but it is improving over time.

7.2.4 Importance of Blockchain Oracles

Data input sources for blockchain networks are known as blockchain oracles [11], as aforementioned blockchain is immutable, data can't be modified once it is added to the chain, so it is very essential to understand and develop robust practices for data acquisition methods. Blockchain is being celebrated for its visibility & transparency, if the source of data input is not trustworthy, even BCT can't help to establish trust. Here the use of IoT devices comes in handy to acquire data from assets without any human intervention, on the other side we have third-party tools to integrate smart contracts with the physical world including geospatial information [12].

7.3 Blockchain Technology in Supply Chains

Blockchain is assumed to be a good fit for digitizing SCs, as both complement each other to their strengths. Supply chains are complex and unique in their own ways, it is required to digitize the supply chain activities, and at the same time, it is difficult to do so. Already many individuals are discussing this, and few are trying to implement BCT into their SC networks, as in current times, visibility of network, information transparency, and activities traceability becoming prominent for stakeholders. Nowadays, there is a natural curiosity in people, to know what they are consuming and purchasing. They expect visibility in a credible manner since fake or counterfeit products are so common in modern times.

Other hand, companies or regulated bodies must incentivize people (suppliers, consumers, manufacturers, etc.) [13] who are promoting good to great practices in respective fields in their industry [14]. So, it is required to gather and maintain information in a transparent way, which helps in appreciating the right people, BCT can facilitate this mechanism. Visibility and transparency, these two terms may look similar, but they have different senses. Visibility provides a sneak peek into the activities, whereas transparency comes along with accountability, which is very much required to ensure a smooth flow of activities. Blockchain brings both into the SC, given that integration of SC and BCT is done right.

7.3.1 Suitability of BCT for Specific Use Case (Including SC)

BCT can couple trust, visibility, transparency, and traceability into any application, given a proper and tighter integration among assets and BCT ledger.

But without a proper use case, BCT complexes the traditional practices. If a traditional database is fulfilling every requirement of stakeholders, there is no need to rush behind a BCT-based database. So, it is required to evaluate BCT against need statements along with the existing information technology practices, with the framework-based evaluation organizations can conclude the kind of BCT suitable for their infrastructure and requirement [15]. Data science can be implemented into practices with smaller scales as well, of course bigger scale yields efficient decision-making, whereas BCT must be started with a reasonable scale and the process is capital intensive. So, there should be good evaluation and strong use cases to achieve return on investments through BCT.

7.4 Opportunities for Digital Technologies in Supply Chain

Adapting digital technologies is trivial for SCs, as they are complex, it is required to make decisions in data-driven and transparent manner. There are opportunities for digital technologies to address in SC [16], those are

1. **Data** needs to be acquired from every step of SC, few suppliers lack machine automation systems to produce data, there is still manual entry of data that keeps happening (It is a threat to BCT), and few locations lack internet services as well. Stakeholders should remember data is a prerequisite for everything, moreover, data acquisition methods should be robust

2. **Storing data in distributed ledgers rather than in centralized ledgers**, centralized ledgers are not so efficient in terms of transparency and traceability. Also, in centralized databases, there is a high chance that data is ending up in silos. Before adding the data, ensuring acquired data quality and conformity

3. **Using data understanding the methods**, process, and mine for specific value-adding insights, which are trivial for visibility

4. **Utilizing** every possible bit of data to establish data-driven decision-making, also using data for predictive modeling to prepare for worst- and best-case scenarios

5. Facilitating **end-to-end operations** through smart contacts.

Through the above steps, companies can develop efficient and robust information systems which promote data-driven decision-making in a transparent manner for supply chains. The above steps deal with higher level tasks, always ground-level reality is different from expectations.

7.5 Challenges for Digital Technologies in Supply Chain

Challenges are part and parcel of any and every task, there are no shortcuts to complete digitization of SCs [17], every step needs to be digitized in a robust and failproof manner. To name a few challenges [11]

1. **Lack of infrastructure**, IoT devices are not fully deployed to gather the data, even if it is installed, ensuring sensors working condition and data transfer is trivial

2. **Lack of complete usage of data**, though less data is available compared to ideal state, still entire data is not being subjected to detailed analysis, data is being used for occasional technical checks and mandatory logs fulfillment in most of the cases.

3. **Interoperability**, BCT will be having a cluster of computers to process the information, as every individual or organization has a decentralized ledger, if a particular system fails, it will be difficult for BCT to maintain the chain without losing out any information. Also, it is difficult to manage the information from a cluster of computers [18]

4. **Lack of BCT development tech skills**, as this field is new and has been in emerging technology for long, many didn't show the interest toward learning BCT skills, now those skills are been picked by individuals in great amount, but finding the right talent is still a challenge

5. **Cost of implantation**, BCT is resource, infrastructure, and capital intensive. Even on a small scale, BCT costs are much higher and the return on investment differs from case to case.

6. **Regulatory bodies** are not up to the mark at this moment, they lack in adopting the latest technologies, even if they do, most of the time they are not reliable

7. **Scalability** of BCT is not as easy as scaling the traditional centralized databases

7.6 Reflection of Society 5.0

From the above discussion, it is evident that Society 5.0 requires technologies like IoT, robotics, data science, machine learning, reinforcement learning, deep learning, AI, BCT, etc., the only difference is all technologies co-exist to form an advanced society. Of course, it will take a lot of time to climb from

Society 4.0 to Society 5.0, still there are underprivileged people, places, and locations, which have not even advanced till Society 4.0. But strategic evaluation of initiatives is required to maintain the equilibrium, technologies are existing to make the lives of human beings better, but not to complex their lives much. An ideal society never aims at snatching work or well-being from humans, including Society 5.0, so, it is wrong to perceive this initiative as a threat to humankind, especially to those who are doing ordinary and mundane jobs. Regulatory bodies and corporations should train people as technology drives, which promotes people and their living to one level up, instead of impacting them. So, upskilling and reskilling will be the gamechanger for stakeholders. Also, there are concerns around eco-friendliness of BCT and crypto mining, as they are consuming a lot of computing power, corresponding authorities are working on it to make them better.

Regulatory bodies and governments are lacking in adopting digital technologies, they are being laggards in BCT. If we wish to see an ideal Society 5.0, the government should be a leader in adopting digital technologies, as humanitarian aspects are driven and controlled by regulators and government bodies. Till now people are so concerned about the supply side of SC; the same efforts should translate to the demand side as well; in humanitarian aspects, the information is key to deliver the essential goods and commodities.

Post any crisis or disaster, governments and companies should focus on delivering the essentials to affected regions. Key to achieve their balance is

1. **Preparedness plan** is required to understand where we stand and what potential steps can be taken to rescue and extend help to needy people. Governments should be prepared for best and worst cases, diversifications are good, investments in biased areas lead to worse consequences. Plan must include cost & time saving, stakeholder analysis, and infrastructure understanding [19]

2. **Assessment and planning** are required to point the most affected crowd to start with, also assessment should suggest the next steps in efficient manner, plan must be efficient and quick to mitigate greater risks [19]

3. **Mobility: Sourcing & gathering**, plan sourcing and sending goods to targets, finding optimized methods are important. Sourcing from in country and international markets, transport, and storage is not easy in uncertain time [19]

4. At last **distribution**, the most important and challenging step [19]

If you observe the above steps, it is evident that data is required at every step to ensure the measures are going as planned and technologies driving the changes. IoT and blockchain oracles help in acquiring the data, BCT will be helpful to maintain and store information, smart contacts will show their presence in transparent decision-making, data science and prediction algorithms are there to foresee the future and be prepared for the worst, of course technologies are very much essential to ensure timely and transparent delivery of intended services. Corporate is leading the game as of now, regulator bodies are lacking in encouraging digital technologies in SC activities. Crisis or disasters won't show a precedent step as indicators; they can occur at any moment or in any manner, so, being prepared all the time is the best thing regulator bodies can do. Adopting digital technologies by setting up the required infrastructure won't happen in a snap; it is a progressive development.

Since adopting digital technologies is new for most of the authorities, they can leverage design thinking frameworks to efficiently embrace digitalization into traditional practices [20]. The steps would be

1. **Empathize**, talk to respective stakeholders, gather their statements, understand the pain points to understand the domain and use case dynamics in detail.

2. **Define**, post empathizing, there will be a clarity on domain and problem space, convert stakeholder pain points to a problem statement.

3. **Ideate** around the defined problem statement, to find the patterns and root causes of the problem. In ideation, practitioners can certainly comment on whether their approach is in right direction or not.

4. **Prototype**, based on the learnings till now, prepares a basic version of conceptualized solution, it need not be full in scale, at least it should be a working example.

5. **Test**, now practitioners will have a minimum viable product in hand, test its functionality and iterate over the steps to get the desirable outcome.

Post implementing the technologies, it is important to automate the previous knowledge as well. The current & future data alone won't speak about everything, one can utilize historical data and measures taken at the time of crisis. So, digitize the historic learnings, compliance, accountability, and standard operating practices.

7.7 Limitations and Future Scope

This study makes a significant effort in providing the context around digital technologies, BCT in SC, and Society 5.0. Moreover, this study is a subjective piece of review, and that's why we highlighted the potential opportunities and challenges in a precautionary manner, also the design thinking framework has been highlighted to solve critical problems in addressing the humanitarian aspect of SC. However, these measures can be considered while attempting to implement the BCT in SC. Limitations of this study include that there can be more pain points that BCT can address and BCT can create in the process of implementation in SC. Also, Society 5.0 is still in its early stage of emergence, so there will be uncertainty around the acceptance and the scale Society 5.0 can achieve in the near to long term, in this study it has been assumed that masses will accept Society 5.0 in a paced manner.

Future scope would be, considering a small chunk of the supply chain to understand how digital technologies adoption can result in efficiency of SC and how return on investment (ROI) looks like. Also, it is up to researchers to understand the role of digital technologies in detail and pave the way for practitioners to brainstorm, practice, and implement technologies in SC.

7.8 Conclusion

Digital technologies have dominance till now, there will be best practices and right use cases for any technology at every instance, stakeholders should find the balance in adopting the technologies for the greater change, design thinking principles are there to ensure right approach. The pace corporates are having in digital technologies adoption, regulator and government bodies should pick up to promote humanitarian aspects of SC. BCT is a good fit for digitizing SC activities, but every use case of SC can't be efficiently carried out by BCT, it requires stringent evaluation to ensure that the use case is suitable, opportunities and challenges for digital technologies in SC aspect are discussed above. There is a tough and long transition from information age to Society 5.0, which is there to ensure economic advancements of people, but we live very far from that as of now. Still the efforts must continue to ensure the technologies are helping human aspects as much as possible; regulatory bodies should lead these efforts.

References

1. Society 5.0 https://www8.cao.go.jp/cstp/english/society5_0/index.html
2. Raj Kumar Reddy, K., &, Kalpana, P. (2021). *"Chapter 11 Impact of COVID-19 on Global Supply Chains and the Role of Digitalization: A VUCA Approach"*. Springer Science and Business Media LLC. https://doi.org/10.1007/978-3-030-72575-4

3. Tiwari, S., Wee, H. M., & Daryanto, Y. (2018). Big data analytics in supply chain management between 2010 and 2016: Insights to industries. *Computers & Industrial Engineering*, 115, 319–330.
4. Yi, H. (2019). Securing instant messaging based on blockchain with machine learning. *Safety Science*, 120, 6–13.
5. Sharma, R., Kamble, S. S., Gunasekaran, A., Kumar, V., & Kumar, A. (2020). A systematic literature review on machine learning applications for sustainable agriculture supply chain performance. *Computers and Operations Research*. (Advance copy)
6. Attaran, M., & Gunasekaran, A. (2019). Blockchain-enabled technology: The emerging technology set to reshape and decentralise many industries. *International Journal of Applied Decision Sciences*, 12(4), 424–444.
7. Alladi, T., Chamola, V., Parizi, R. M., & Choo, K. R. (2019). Blockchain applications for industry 4.0 and industrial IoT: A review. *IEEE Access*, 7, 176935–176951.
8. Humanitarian supply chain. http://www.helplogistics.com/
9. Jaoude, J. A., & Saade, R. G. (2019). Blockchain applications – Usage in different domains. *IEEE Access*, 7, 45360–45381.
10. Raj Kumar Reddy, K., Gunasekaran, A., Kalpana, P., Raja Sreedharan, V., & Arvind Kumar, S. (2021). Developing a blockchain framework for the automotive supply chain: A systematic review. *Computers & Industrial Engineering*, 157, 107334. ISSN: 0360–8352. https://doi.org/10.1016/j.cie.2021.107334
11. Dolgui, A., Ivanov, D., Patryshev, S., Sokolov, B., Ivanova, M., & Werner, F. (2020). Blockchain-oriented dynamic modeling of smart contract design and execution in the supply chain. *International Journal of Production Research*, 58(7), 2184–2199.
12. Viriyasitavat, W., Anuphaptrirong, T., & Hoonsopon, D. (2019). When Blockchain meets the internet of things: Characteristics, challenges, and business opportunities. *Journal of Industrial Information Integration*, 15, 21–28.
13. Nayak, G., & Dhaigude, A. S. (2019). A conceptual model of sustainable supply chain management in small and medium enterprises using blockchain technology. *Cogent Economics and Finance*, 7(1), 1667184.
14. Blackman, I. D., Holland, C. P., & Westcott, T. (2013). Motorola's global financial supply chain strategy. *Supply Chain Management*, 18(2), 132–147.
15. Fraga-Lamas, P., & Fern'andez-Caram'es, T. M. (2019). A review on blockchain technologies for an advanced and cyber-resilient automotive industry. *IEEE Access*, 7, 17578–17598.
16. Ivanov, D., Dolgui, A., & Sokolov, B. (2019). The impact of digital technology and industry 4.0 on the ripple effect and supply chain risk analytics. *International Journal of Production Research*, 57(3), 829–846.
17. Xu, X., Jin, Y., Zeng, Z., Yang, S., & Chen, R. (2019). Hierarchical lightweight high-throughput Blockchain for industrial internet data security. *Computer Integrated Manufacturing Systems, CIMS*, 25(12), 3258–3266.
18. Zhao, G., Liu, S., Lopez, C., Lu, H., Elgueta, S., Chen, H., & Boshkoska, B. M. (2019). Blockchain technology in agri-food value chain management: A synthesis of applications, challenges and future research directions. *Computers in Industry*, 109, 83–99.
19. Design thinking. https://designthinking.ideo.com/
20. Bai, C., & Sarkis, J. (2020). A supply chain transparency and sustainability technology appraisal model for blockchain technology. *International Journal of Production Research*, 58(7), 2142–2162.

8

Opportunities and Challenges for Society 5.0 in Supply Chain Management

Abdul Zubar Hameed, Mohammad Hani Kashif,
Yazeed Abdul-Monem Al-Mahi, Salman Ali Al-oufi, and K. Lenin

CONTENTS

8.1 Introduction

Society 5.0 can be defined as "a people-centered society that equalizations monetary advancement and social issue fathoming through a system that firmly coordinative the web world and therefore the physical space". The 5th Technology and Science plan see Society 5.0 because of the long-haul society that Japan needs to purpose for. Look Affiliations are separated into four bunches: Society 1.0 Hunting & gathering, Society 2.0 Agricultural, Society 3.0 Industry, and Society 4.0 Information.

Exchanging knowledge and information across departments wasn't enough in Society 4.0 (Information), also cooperation was difficult. Due to the limited potential of humans, finding and evaluating critical information from large amounts of data has been tedious, limiting the scope of work and behaviour by age and talent. There were also various restrictions to address concerns such as lower birth rates, ageing populations, and depopulation of the region, making it difficult to respond appropriately. Social reform in Society 5.0 will result in a future-oriented society that breaks out of its current rut, a society

DOI: 10.1201/9781003177432-11

where members respect one another across generations, and a society where everyone can actively participate in having fun (Figure 8.1).

The main objectives of this chapter are to identify the Society 5.0 terminology and how it works as well as summarizes its history and identifies the main opportunities and challenges while applying Society 5.0 terminology in supply chain management (SCM). To achieve these objectives, we need to ask some sub-related questions such as what Society 5.0 is, what are the functions of Society 5.0, what is SCM, and how does it relate to Society 5.0. Finally, what are the opportunities and challenges of Society 5.0 in SCM?

8.2 How Society 5.0 Functions?

Actual space & Cyberspace (virtual space) are extensively intertwined in Society 5.0 (Real space). People used to use cyberspace cloud services (databases) through the Internet to search, retrieve, and analyse information and data in today's information society (Society 4.0). The Society 5.0 Internet, on the other hand, collects massive amounts of data from sensors in physical space. Artificial intelligence (AI) analyzes these huge amounts of data in cyberspace, and the results will be at humans in physical real life in several formats. To this point, people in the information society have acquired and accessed data via the Internet. In Society 5.0, people, things, and systems are all linked together in cyberspace, and the finest AI results that go beyond human capabilities are transferred into physical space. This method delivers importance to the industry and society in previously unthinkable ways (Figures 8.2 and 8.3).

8.3 Supply Chain Management

SCM can be defined as the management of the transportation of goods and services, and it encompasses all operations that turn raw materials into completed commodities. It is the proactive simplification of a company's supply-side process to maximize customer profitability and gain a competitive advantage in the market. SCM is the central management of the flow of products and services, and it encompasses the full process of converting raw materials into finished items. Companies can cut expenses and deliver items to customers faster by simplifying their supply chains. Companies benefit from SCM because it protects them from disruption, costly recalls, and litigation. Planning, raw material procurement, production, distribution, and returns are the five most important aspects of SCM. Managing and reducing costs and avoiding delivery bottlenecks are SCM tasks.

Society 5.0

Society 1.0
Hunting & Gathering

Society 2.0
Agricultural

Society 3.0
Industrial

Society 4.0
Information

Society 5.0

FIGURE 8.1
What is Society 5.0?

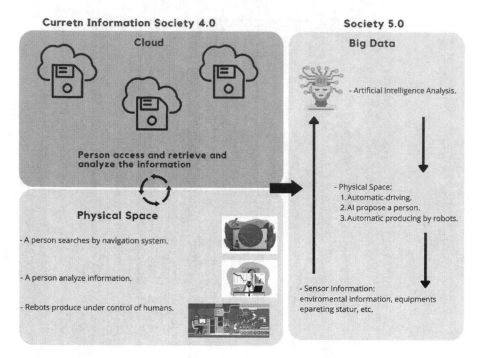

FIGURE 8.2
How Society 5.0 works?

Through SCM, suppliers aim to create and manage the most efficient and cost-effective supply chain possible. Everything from production to product development, as well as the information systems required to manage these activities, is included in the supply chain. SCM is typically intended to centralize or link the manufacture, shipment, and distribution of products. Companies can cut expenses and deliver items to customers faster by simplifying their supply chains. Internal inventory, internal manufacturing, distribution, sales, and inventory held by firm suppliers are all being closely controlled. SCM is founded on the idea that nearly all products on the market came from numerous enterprises working together to construct the supply chain. Although supply chains have been around for a long time, most businesses have only recently recognized them as a significant addition to their operations. This chapter describes the opportunities and challenges of Society 5.0 in SCM[1].

8.4 Literature Review

Although Society 5.0 is quite a new topic in the spectrum, there have been numerous journals and papers published in the literature. So, to better introduce

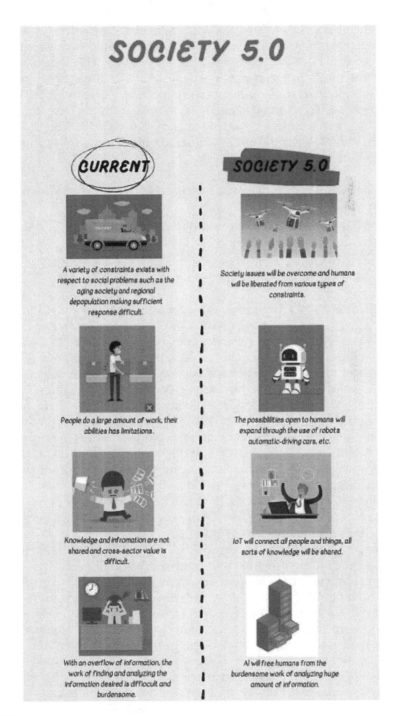

FIGURE 8.3
Achieving Society 5.0.

the concept of Society 5.0 in supply chain and logistics, we will go over the published literature on related topics such as industry 4.0, and supply chain 5.0.

The first paper titled "Industry 4.0 and Society 5.0: Opportunities and Threats" discusses and analyses the openings and challenges behind industry 4.0, the industrial transformation introduced by Germany in 2011, as well as Society 5.0, a promise that began in Japan to reshape society, and to better understand the connection between them and the contributions of each concept to the other[2]. As for the models or paradigms of industry 4.0, the authors define them as follows:

- "Intelligent Product, where data related to operation cand standards can be saved".
- "Intelligent Machine, where self-organizing behaviour is prevalent instead of traditional methods".
- "Augmented Operator, where flexible parts of the process are possible through automation knowledge".

As for the main founding parts, the cyber-physical system (CPS), a computer system controlled by algorithms, the Internet of Service (IOS), which describes the availability of software and software creating tools online, as well as the Internet of Things (IoT), which involves the concept of physical objects connected to the Internet via sensors and exchange data with other connected objects. The previous parts go on to back an expression called smart factories, where factories can execute commands and share data.

Throughout the various connected devices. Industry 4.0, however, does come with its shares of challenges, namely competitiveness in implementing the smart solutions of the industry with other organizations in quality, delivery time, and efficiency, all while managing the high costs of implementing such technologies. Furthermore, behind industry 4.0 lies the need to implement different complex systems, such as CPS, and integrate them with the existing processes, with the need for highly skilled technical personnel to guide the implementation[3].

With regard to Society 5.0, it can be described that it is a human-focused technological evolution that aims to improve the quality and well-being of the people it surrounds. This plays hand in hand with Japan's ageing society and promotes the use of technologies like autonomous vehicles to automate basic life activities. A well-known technology concept in Society 5.0 is Parallel intelligence, defined as the connection between virtual and actual reality, which is regarded as the end-stage of mechanization and electrification. Another concept is smart societies, where values are interconnected through the CPS and intelligent societies, as IoT to improve all aspects of life, rather than a sole focus on the industrial sector as in industry 4.0. It demonstrates how problems such as ageing and natural disasters through the integration of technology with society.

These days, the supply chain and logistics depend on humans' capabilities as well as robots. The concept of technology expanded so fast. In addition to robots, supply chain 4.0 uses new technologies such as the IoT, Cloud Computing, Additive manufacturing, and Blockchain. All the mentioned technologies are effectively integrated and allow better performance in the supply chain. The main aim of using this kind of technology is to create a CPS overall the logistics processes[4].

The first time, Society 5.0 terminology has been mentioned in January 2016 by the Japanese government. Society 5.0 predicts the intensive use of current industry technologies going beyond the manufacturing and business perspective. However, to implement Society 5.0 ideas, the current supply chain needs to develop forward, supply chain 5.0 needs more collaboration between humans and robots. The main objective of using robots is to be more focused on the roles and jobs that are harder and more dangerous for humans (Figures 8.4 and 8.5).

Financial capital is not the maximum vital element in Society 5.0; instead, information. The IoT will herald a brand new technology committed to understanding human potential. These studies intend to investigate the feasibility and issues of imposing Fintech in small- and medium-sized corporations. A survey of 60 small- and medium-sized corporations was carried out with this intention. Due to regulations in technical infrastructure, information and information, human resources, and regulation, the application of Fintech in medium industries and small businesses did now no longer seems geared up to be realized. As a result, those factors have to be progressed so one can gain a human-cantered and technology-primarily based society.

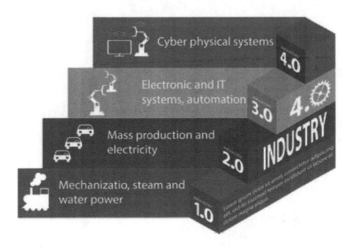

FIGURE 8.4
The digital supply chain 4.0.

FIGURE 8.5
The society development over centuries.

In Society 5.0. Fintech will consider a large development in the financial sector. Fintech can be considered partly driven by the sharing economy and technological advances. The main aims of Fintech are to provide a new low-cost financial structure, improve the quality of service in the financial sector as well as to provide a broad and stable financial environment. Luckily, due to the large development in the infrastructure during the last decade as well as developing the smartphone technologies and increased investments in big data, financial start-ups can revolutionize the financial sector with innovations, and personalized & customized services. Currently, Fintech left the hyper stage and represented itself as a major player in the financial sector. The above benefits and others distinguish Fintech from traditional financial companies. Additionally, Fintech provides a competitive advantage to start-ups in the financial sector. Several financial companies started to take Fintech start-ups seriously; they begin to adopt new strategies and methods to compete with them.

According to several studies, the human factor has a high positive impact and effect on FinTech. This suggests the financial companies' staff need to be reliable and qualified. These staff will demonstrate scientific computing skills and provide services tailored to the needs and characteristics of their clients. Indonesia's digital economy is constrained by a shortage of talent. Therefore, it is up to the university to solve this problem. The results of this study suggest that the current regulatory and government policies do not appear to support the use of Fintech in SMEs. The factor loadings for every single indicator in the PLS system show this. Therefore, these factors need to be improved to achieve Society 5.0.

In the previous papers, the researchers tried to identify the concept of Society 5.0 in general; here we will try to identify the concept of Society 5.0 in detail. Additionally, we will explain the effect of Society 5.0 on the supply chain. Also, the main opportunities and challenges of implementing Society 5.0 terminology in SCM practices. Previous papers also greatly expanded on technical terms associated with Society 5.0 such as IOS, as well the challenges of implementing similar concepts such as industry 4.0 in the various industries. In addition, previous papers have studied the effect of Fintech (Financial technology) on small- and medium-sized companies and the relationship with Society 5.0 (how we can achieve Society 5.0 via Fintech).

8.5 Opportunities for Society 5.0

Society 5.0 can be regarded as a human-driven technological evolution that favorite financial growth and tackles social challenges, such as ageing, by harnessing technologies such as IoT and IOS. Supply chains are the sets of networks and activities that involve companies and people, that aim to provide a product or service. In hindsight, these two concepts don't seem related nor intertwined, but Society 5.0 can help in coordination, information sharing, among other areas sestina to the supply chain. In the following few paragraphs, we will uncover the various opportunities of Society 5.0 in the supply chain spectrum.

1. Coordination

Data and information are crucial aspects of the coordination in supply chains, be it the current inventory level, the location of the shipment on the road, or in the warehouse. IoT, an innovation part of the Society 5.0 technologies, refers to connecting numerous data points and is widely used in supply chains, to provide streams of real-time data for better decision making and rapid response to changing events. A few IoT applications are RFIDs, which track location in real-time and inform management of deviations in direction prior in time, storage condition sensors which inform us of changes in environmental factors such as temperature and humidity, or location tags which help minimize the time looking for specific items. Overall it helps with coordination, optimization, and decision making through the feature of live information sharing.

2. Labour saving

An emerging concept in the field is the automounts supply chain, a chain that combines standardization, interconnectedness, and intelligence, where the chain can process orders, gather the location of the item, and deliver the unit. All in all, a concept that minimizes the input needed from humans to fully function. This concept greatly affects the human factor as it can reduce the load on manual labour, saves time by eliminating repetitive non-value-adding tasks, lowers overhead, increases agility with free-flowing information which helps the supply chain team deal with rapid changes, and can also decrease the natural human-related errors in processes and tasks.

3. Social contributions

Supply chains involve plenty of modes of transportation, be it cargo airplanes, large trucks, or container ships, all methods that use large amounts of fossil fuel and produce toxic, greenhouse gases and emissions that trap heat in our atmosphere, which can then contribute to global warming, among other environmental disasters. To combat such long-lasting adverse effects, Society 5.0 promotes the use of next-generation vehicles such as electric vehicles, and liquified natural gas-fuelled ships, which can release 45%–50% less fuel than coal and less oc2 than oil by 30%[5].

8.6 Challenges for Society 5.0

Although Society 5.0 is a new thermology in the spectrum, there have been several journals and research papers published about it in the last decade, since Japan government promoted Society 5.0 terminology, several challenges appear on the side of SCM. Terminologies such as Society 5.0 require the government, companies, and local people to change their business, re-engineer some major processes and even change the lifestyle of people. Implementing Society 5.0 terminology requires the government and private sector to build a long-term plan (between 15 and 20 years durations). There are several possible challenges in front of any counties wants to implement Society 5.0 terminology. For example, but not limited to:

1. Cyber Threats

    Although cyber security developed during the last decade. The cyber threats increased parallelly, especially during the last 2 years (during the COVID-19 pandemic). Society

    5.0 terminology requires us to increase the use of technologies and incorporated them into every corner of our social being, and our physical being. Subsequently, the risk to our personal and collective safety will increase. In addition, public safety will be negatively affected since autonomous vehicles, AI-operated public transport systems, fleets of drones, smart grids, and connected medical devices can all be hacked. According to some statistics, during 2020 there were about 26,000 cyber threats per day with a total number of 9.66 million during the year. Most of the cyber threats and attacks happened in the USA, UK, France, South Korea, China, and Japan. Nearly half of cyber-attacks targeted public services such as public transportation systems, health care systems, and other services. In the current days, applying all parts of Society 5.0 thermology will threaten public safety and the stability of the country. Fortunately, most first-world countries started investing and paying attention to cyber security & computer security. For example, in the United State, the total capital investment in cyber security increased from 6.9 Billion US Dollars in 2020 to 17.4 Billion Dollars in 2021. In addition, in 2020 Japan government spend over 25.6 Billion JP¥ 2020. All this kind of information indicates that cyber threats & attacks will reach a block point in the future.

2. Infrastructure

    Society 5.0 terminology requires integration with the country's infrastructure. Society 5.0 revolves around high-speed broadband connections and ubiquitous mobility. These types of technologies request the local government to re-engineer network facilities.

Additionally, to apply Society 5.0, many countries require to add new networks systems, new facilities, and new regulations to their infrastructure, Subsequently, will cost billions of dollars and years of hard work. Currently, some counties started to develop their infrastructure to match Society 5.0 terminology. The best example is Japan, during 2019, the Japanese government spends over 1.44 Billion JP¥ to develop AI & big data technologies, and their plan to grow to 4 Billion JP¥ in 2034. On the robots' side, they spend almost 6 JP¥ during the last decade to develop and expand the employment of robots. On the other side of the map, many European counties started investments in AI technologies such as France with a total investment of 1.5 billion euros in 2022, and the government plans to increase its investment to 2.2 billion euros by the year 2025. Through all these kinds of investments and plans, the Society 5.0 terminology was possible to achieve during this century.

In the end, several other challenges face Society 5.0 terminology technical challenges, materials limitations, security confrontations, and even financial challenges. Large scale terminologies such as Society 5.0, require a long-term plan as well as a short-term, in addition to human competencies with long-term investments from the government sector as well as from the private sector.

8.7 Discussion

This chapter talked about Society 5.0 and SCM and the opportunities and challenges that Society 5.0 faces in SCM. We started by defining Society 5.0 and SCM to give a clear picture. After that we talked about three opportunities that Society 5.0 faces in SCM and they are as follows: Coordination, Society 5.0 can be applied to SCM through the feature of live information sharing. Labour-saving, this concept has a significant impact on the human factor because it reduces the burden on manual labour, saves time by eliminating repetitive non-value-adding tasks, lowers overhead, increases agility with free-flowing information that assists the supply chain team in dealing with rapid changes, and reduces natural human-related errors in processes and tasks. Social Contributions, Society 5.0 encourages the adoption of next-generation vehicles such as electric vehicles and ships powered by liquefied natural gas, which emit 45%–50% less fuel than coal and less oc2 than oil by 30%. Furthermore, we identified two challenges that Society 5.0 faces in SCM, cyber threats, the term "society 5.0" refers to a growth in the usage of technology that is infused into every aspect of our social and physical selves. As a result, the risk to our individual and communal safety will rise. Furthermore, public safety will be jeopardized since autonomous vehicles,

AI-controlled public transportation systems, drone fleets, smart grids, and connected medical gadgets can all be hacked. Infrastructure, the concept of Society 5.0 necessitates a connection to the country's infrastructure. High-speed broadband connections and ubiquitous mobility are fundamental to Society 5.0; urban preparedness also encourages intense construction procedures for many forms of IT infrastructure. These technologies necessitate the reengineering of network infrastructure by the local government. In addition, many countries will need to add new network systems, facilities, and rules to their infrastructure to implement Society 5.0. As a result, billions of dollars and years of effort will be required.

8.8 Conclusion

Terminologies such as Society 5.0 will change the lifestyle of human beings for the better. Society 5.0 terminology will help to eliminate the regional, age, gender, and language gaps. Society 5.0 will be a large improvement in the lifestyle of all people. As known, large improvements require large changes and lots of effort from governments global companies, and local businesses. This chapter starts with a short Literature-view about Society 5.0 and SCM. The main body of this chapter is opportunities for Society 5.0 in SCM and challenges for Society 5.0 in the supply chain. Finally, this chapter ends with a short discussion to summarize the main ideas and general information

References

1. Mofokeng, T.; Chinomona, R. 2019. Supply chain partnership, supply chain collaboration, and supply chain integration as the antecedents of supply chain performance. *S. Afr. J. Bus. Manag.*, 50, 30–58.
2. Pereira, A. G.; Lima, T. M.; Charrua-Santos, F. 2020. Industry 4.0 and Society 5.0: Opportunities and threats. *Int. J. Recent Technol. Eng.*, 8(5), 3305–3308. https://doi.org/10.35940/ijrte.d8764.018520
3. Frederica, G., 2021. From supply chain 4.0 to supply chain 5.0: Findings from a systematic literature review and research directions. *Logistics*, 5(3), 49.
4. Dolgui, A., Ivanov, D.; Sethi, S. P.; Sokolov, B. 2019 Scheduling in production, supply chain and Industry 4.0 systems by optimal control: Fundamentals, state-of-the-art and applications. *Int. J. Prod. Res.*, 57(2), 411–432. https://doi.org/10.1080/00207543.2018.1442948
5. Wilkesmann, M.; Wilkesmann, U. 2018. Industry 4.0—Organizing routines or innovations? *VINE J. Inf. Knowl. Manga. Syst.*, 48(2), 238–254.

Section 4

Industry Application

9

Blockchain Application in Cryptocurrency

Atour Taghipour, Basma Addakiri, and XiaoWen Lu

CONTENTS

9.1 Introduction

The asset management and investing world are always changing and evolving. In a highly competitive industry, fund operators, corporations, and consumers are continuously looking for the most effective approach to differentiate themselves. Blockchains have been coupled with artificial intelligence and big data, but this is an ever-changing landscape (Hassani et al., 2018). During the financial crisis, we learned that various securitization strategies can be extremely destructive to financial markets and the general public if proper due diligence and monitoring are not performed (Castelluccio, 2018). Between all actors in the securitization chain, there is a lack of comprehensive transparency. This has an impact on the underlying assets' audit and rating. While new technologies have taken their place, they are beginning to provide additional benefits. Asset management and investment have traditionally been very institutional and private, with their own set of rules and procedures. With the advent of blockchain, artificial intelligence, big data, and a variety of other technological tools, everything is changing. More importantly, the study of blockchain segmentation appears to offer a fresh perspective on asset management and investment. On the other hand, asset tokenization is similar to a securitization operation, which involves the

DOI: 10.1201/9781003177432-13

division of financiers, the appearance, and partial ownership of assets sold. Indeed, bankers securitized and sold trillions of dollars in mortgages without regard for the loan's potential risk profile, resulting in the 2008 financial crisis. This is why the blockchain is so essential in this type of securitization, since security tokens, which are becoming increasingly popular in the form of STO "Secure Token Products", benefit from the lack of mediators. However, in this new world of asset investment, there are also other issues and risks to be aware of. When Bitcoin was launched, the blockchain was created to allow users to exchange the same-named cryptocurrency (Davidson et al., 2015). Following that, a slew of blockchains popped up. The most well-known, Ethereum, is praised for its capacity to install and interact with smart contracts (Pustišek and Kos, 2018). A smart contract is computer software engrained in a block that keeps running when special circumstances or occurrences are satisfied. This requires that the blockchain system should be able to gather information from outside the blockchain that ensures the contract's terms and conditions are met (Szabo, 1997). Smart contracts for the blockchain enable not only the execution of functions but also the storage of states. They benefit directly from the blockchain's unique qualities, including as integrity, decentralization, lack of transaction repudiation, and transparency. Unlike traditional trusted third parties that verify the legitimacy of transactions through their status, such as banks or governments, the blockchain has the status of an artificial trusted third party, where the code and power of the network may be trusted. These features have piqued the interest of business leaders and academics, who view the blockchain as a way to transform the way people exchange value while also ensuring the validity and integrity of the data stored on it. Indeed, the blockchain would be "a native digital medium for value, via which we might manage, store, and exchange different assets, peer-to-peer and securely", according to the report (Tapscott, 2016). In this chapter, we first study the link between blockchain and cryptocurrency, next we develop a new discussion about the technological evolution of the currency due to blockchain.

9.2 Literature Review and Literature Gap

In many sectors, such as supply chain management, banking, network control, digital decentralized identification, and healthcare, there are many applicable blockchain use examples in the literature (Andreev et al., 2018). Despite its potential, however, blockchain has a multitude of barriers in terms of acceptance. According to a survey done by the firm PwC in 2018, businesses encounter issues such as ambiguous regulation. Due to the collaboration of the parties in the blockchain ecosystem as well as the necessary legal and political entities, these issues are gradually being overcome. Nonetheless, for a variety of reasons, businesses continue to encounter a technological hurdle. They may find it difficult to recruit employees specialized in blockchain, as the technology is still young. They may also find it difficult to integrate blockchain into their existing information systems and business processes, as there are no best practices identified and proven in companies by software architects. Considering blockchain is still a developing technology, they may have a hard time finding individuals who are experts in the field. They may also find it challenging to integrate blockchain into their existing information systems and business processes because software architects have yet to identify and validate best practices in the field (Risius and Spohrer, 2017).

9.2.1 Blockchain and Cryptocurrency

9.2.1.1 Blockchain

Blockchain is a method of storing and sending data that works without the need for a central administrator to oversee the entire system. It tries to connect computers without the use of a central database, with servers picked at random based on their computational capacity; "miners" approve the blocks in exchange for payment, such as Bitcoin. For trading, this system relies on the use of asymmetric cryptographic keys and hashing methods (Andriole, 2020). These procedures assure transaction integrity and authentication. Indeed, the operation's reliability is confirmed using the public key that matches the private key used to sign the transaction. As a result, each user can verify the validity of transactions on a public, anonymous, and secure digital register that contains a distributed database and is hosted in a network of secure computer stations. This database stores the history of all exchanges and allows users to share it. Electronic time-stamping techniques are used to keep track of transactions. Each block comprising a series of transactions is linked to the next higher version of the block, and each transaction is checked chronologically before being integrated. Each block that contains a set of transactions is linked to the next block (Figure 9.1). Each transaction is confirmed in order and then included in the block's next higher version. However, because the system's security relies on cryptographic techniques, it's critical to have adequate levels of reliability (Andriole, 2020). A blockchain's security could be damaged otherwise. Furthermore, the blockchain contains a number of smart contracts, which are programs that interact with one another, are accessible and auditable by all authorized parties, have controlled and verifiable execution, and are designed to automatically execute the terms of a contract when certain conditions are met (Von Haller Gronbaek, 2016).

Blockchain technology is generally characterized by:

- *Disintermediation*: Blockchain technology provides for the interchange of information and valuables, as well as the control and validation of actions undertaken without the intervention of a central

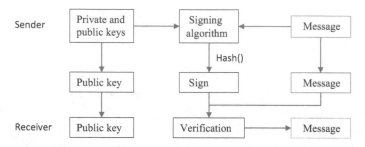

FIGURE 9.1
Transaction in blockchain scheme.

Public blockchain	Private blockchain
Decentralized system	Semi decentralized system
Everyone participation	Limited participation
High output	Low output

FIGURE 9.2
Public and private blockchain.

authority, implying that the blockchain is decentralized and therefore the trust is distributed.

- *Transparency*: The blockchain is transparent because anyone may download it in its entirety at any time and check its integrity. Everyone may observe the current and previous transactions and exchanges, allowing them to verify the chain's legitimacy.

- *Security*: Because the data is not stored on a single server but rather on the computers of some of the users, it is hard to destroy all copies of the papers.

- *Autonomy*: The network nodes, or users, provide computing power and hosting space. As a result, there is no need for a centralized infrastructure because it is shared by all users.

There are two sorts of blockchains: public and private (Figure 9.2).

- The *public Blockchain* is the "conventional" blockchain. That is, a blockchain that anybody in the world can read, and to which anyone can send transactions and expect them to be included in the registry, at least as long as those transactions follow the blockchain's rules. This is the case with Bitcoin, for instance, where anyone can use it for free. This is the same with Bitcoin, for example, where the registry is open to all. Everyone is also free to participate in the approval process, which determines which blocks will be added to the chain and defines the system's present status. Cryptoeconomics, or the combination of economic incentives and cryptographic verification techniques, secures public blockchains as a replacement for centralized networks (Allen et al., 2017). To put it another way, as with Bitcoin, everyone has an economic incentive to participate in transaction verification.

- It is possible to deploy blockchains that are exclusively available to specific individuals, and these blockchains are referred to as *private blockchains*.

Specific actors can have access to a variety of rights:

- When transaction approval is centralized at a single actor, the use of a blockchain is no longer relevant. It's equally as efficient to use a centralized database.
- Transactions are carried out.
- The transaction consultation and the formation of private blockchains necessitate the establishment of a dedicated infrastructure within businesses, as well as the definition of a governance mechanism and server maintenance. The return on investment of such initiatives is frequently disappointing as a result of these investments. Many players have discovered that public blockchains can help them innovate more, for example, by allowing them to use cryptocurrencies or tokenize some of their assets.

9.2.1.2 Blockchain and Cryptocurrency

The blockchain was first created as a way of accounting for the virtual currency Bitcoin (Brito, 2013). The same technology is currently being used in a variety of corporate processes, including transaction verification. Permanent records that cannot be altered are established with this technology, enabling for validity to be confirmed by a whole community rather than a single centralized authority. The terms cryptography and currency are combined to form the term cryptocurrency (Cheung et al., 2015). Cryptography is the skill of writing codes, whereas currency refers to a country's monetary system. Cryptography is used to create a virtual and digital currency (Heeks, 2009). Its immaterial aspect is only used for online currency. The blockchain was originally designed to keep track of the virtual money Bitcoin. Currently, the same technology is employed in a range of corporate procedures, including transaction verification (Taghipour, 2009). According to TanxinBlog, with this technology, permanent records that cannot be altered are created, allowing legitimacy to be certified by the entire community rather than a single centralized authority (Ren et Al., 2016). The terms cryptography and currency are combined to form the term cryptocurrency. Cryptography is the skill of writing codes, whereas currency refers to a country's monetary system.

Cryptography has resulted in the development of virtual and digital currency. Because of its immaterial nature, it is only utilized for online currency. Cryptocurrencies differ from regular currencies in that they are acquired through software and computer programs known as algorithms rather than by a government issuing them. Unlike other currencies like the dollar and euro, the value of the yen is not controlled by market forces (Mannaro et al., 2017). However, speculation is equally important in determining the value of cryptocurrencies. In the TanxinBlog, they mention that traditional currencies, which require hard work to obtain, cryptocurrency are obtained through a process known as mining (Taghipour et al., 2022a). The most popular cryptocurrencies are Bitcoin, Ethereum, Litecoin, Monero, and Dash. However, blockchain and cryptocurrency have several commonalities, such as:

- They're both virtual and intangible.
- Blockchains and cryptocurrency are two examples of recent technology advancements. The first blockchain was created in the late 2000s, following the breakthrough of gurus known as Satoshi Nakamoto.
- Cryptocurrencies and blockchains are mutually dependent. Cryptocurrencies are the real tools exchanged, while blockchains give the path of transaction records.

As a result, there are several distinctions between blockchain and cryptocurrency, such as the following:

- A blockchain is a decentralized system that records crypto transactions.
- A cryptocurrency is a virtual tool that is utilized in block transactions.
- Cryptocurrencies can be used for transactions, investments, and wealth storage.
- A blockchain is a system for storing and processing cryptocurrency transactions.
- Cryptocurrencies have monetary value and can be used as a measure of wealth. Blockchains have no monetary value and cannot be used as a measure of wealth.
- Cryptocurrencies such as Bitcoin can be transferred from one account to another. Blockchains are not mobile.

9.2.2 The Application of Blockchain

9.2.2.1 Field of Application

We'll try to describe what blockchain technology can do in terms of applications in this part, based on various use cases. Then we'll try to figure out what the most common forms of theoretical blockchain applications are.

Blockchain can be applied in a variety of sectors; in this section, I'll mention a few of them.

9.2.2.1.1 Smart Contract

A smart contract is a type of contract that is used. Let's go over the basics of a smart contract. It's a piece of software that's tied to a blockchain account. This code contains specific instructions and routines that enable the creation of operating rules. When someone sends a message to the account associated with the computer code, for example, the latter is performed with the message sent as a parameter.

By filling out an online questionnaire, the startup *Otonomos* assists companies commencing the blockchain journey in forming a legal entity in Singapore. The holder(s) then own the entity created using a cryptographic address and a private and public key. As a result, Otonomos provides a peer-to-peer (P2P) administration system for the company's shares. Smart contracts can also be used to set governance rules in this system (Herbert and Litchfield, 2015).

On the other hand, the startup Symbiont has developed a smart contract-based solution for managing financial transactions automatically. For example, once a security is issued on Symbiont's register, it becomes self-contained, eliminating the need for middle and back-office operations. The business, which has teamed up with Gemalto to safeguard and encrypt transaction data, can now automate many of the manual processes still in use in the financial sector.

9.2.2.1.2 Trading Platforms

Bit-Shares is a DAC (Decentralized Autonomous Company) that facilitates the exchange of financial assets between potential buyers and sellers. The assets exchanged on the markets dedicated to these exchanges have values that fluctuate in accordance with the real financial markets, according to the BitShares principle.

Let's look at an example. A Bit-Asset is a type of asset exchanged on the Bit-Shares system. One US Dollar, for example, is represented by the Bit-USD. A Bit-USD is not the same as a USD. It's a cryptocurrency whose price fluctuates at the same rate as the reference value (here the USD) (Yen, 2021). The price of Bit-Gold is determined by the price of gold, and the price of Bit-Silver is determined by the price of silver (Ametrano, 2016).

As a result, Bit-Assets are made-up financial assets that correlate to BitShares, the cryptocurrency that encompasses all assets. BitShares allows you to purchase a variety of Bit-Assets. The purpose is to speculate on assets in order to increase the number of Bit-Assets or BitShares in circulation. If enough people utilize Bit-Assets, the money (Bit-Share) might be used in real-world transactions, similar to Bitcoin.

9.2.2.1.3 Blockchain in IoT

Connected objects are referred to as IOT (Internet of Things). Connected things are, for the most part, modern technologies whose data is centralized and communicated to an authority, which is usually the publisher of the device in question. As a result, this approach contradicts the blockchain's theoretical concept. To provide security and traceability, the activity of these linked things might be kept directly in a blockchain. Then it would just be a matter of retrieving and utilizing the data via customized interfaces (Mougayar, 2016).

On the other hand, the integration of data from connected objects in a blockchain could ensure the confidentiality of object users. Indeed, even if on a public blockchain, all the data is public, the identity of a third-party account is encrypted and therefore the identity of a person is impossible to determine. The visible data that is actually broadcast will be chosen by the user of the object (Lebosse et al., 2017).

For example, IBM has imagined building a washing machine capable of establishing its need for detergent in complete autonomy, and of contacting the nearest supplier in order to replenish its supplies (Taghipour et al., 2022b). This requires a contract between the supplier and the machine itself. Two major points allow the autonomy of the machine: decision-making and information exchange. This can be managed by smart contracts. IBM also has a complete platform to link IoT to a blockchain, this service is called IoT Watson blockchain (Mbiatem et al., 2018).

9.2.2.1.4 Network Infrastructure

This type of blockchain application is the basic foundation for using the blockchain process. One example, which I mentioned in the first section, is Ethereum. Ethereum is a public blockchain that can integrate smart contracts. Ethereum also has its own currency, called Ether. The first block of the chain was created in 2015 and sold 60,000,000 Ether as a pre-sale, and 12,000,000 Ether to programmers. All subsequent Ethers created will be via the mining process. Ethereum allows its customers to use a platform to carry out exchanges, notably with the help of smart contracts, which automate the way transactions are managed (Tliche et al., 2019).

Counterparty is an initiative that resembles Ethereum in that it offers a platform for creating assets, has its own currency, and allows the implementation of smart contracts in the chain it uses: the Bitcoin blockchain. Thus, Counterparty is an alternative to the platform used by Ethereum, but uses the Bitcoin blockchain (Back, 2002).

9.2.2.2 Application Factors of Blockchain

9.2.2.2.1 Unforgeable Data

Blockchain technology is based on the principle of proof of work, which guarantees the generation of a new block of transactions, which in theory cannot be modified. This is one of the most important aspects of this technology since it ensures that the security of the data is not compromised when a block is registered on a blockchain. The integrity of the data is thus ensured and the users of the blockchain do not have any doubts about the level of reliability of the data. As an example, we can imagine a "patent filing" type of transaction, linked to the person who owns it. The action of depositing one's patent on the blockchain provides non-refutable proof of the patent's ownership. Moreover, the patent would hypothetically no longer have to be filed with a public body or institution, or at least the proof of ownership could be justified by the presence of this transaction on the blockchain. An additional benefit is the prevention of fraud (Toyoda et al., 2017). One could also imagine a malicious accountant wanting to change the account lines of the company he works for. However, storing a company's financial documents in the blockchain would ensure unalterable and visible account books. Thus, the use of blockchain in accounting could alleviate some of the problems of fraud.

9.2.2.2.2 Security by Cryptography

The issue of security and personal data is a central question for many people. In order to secure the privacy and data of each user, blockchain-based applications can enable more advanced services than most traditional applications. Let's take the case of a mobile application. Typically, when installing an application, personal data is at the heart of the process: When accepting the installation terms: ownership of the data is transferred to the application publisher, and permissions that allow the publisher to use the data are granted. Throughout the application's lifecycle, the data is not traceable by the user. In a blockchain application, data usage would be different: Users can control all their data, and access it. Users can be the owners of their data (Gao et al., 2018). The user has the ability to give or take back permission to access their personal data. The identity of the users is protected by the private and public key systems explained in the first part (Miraz et al., 2018).

9.2.2.2.3 Reducing Transaction Costs

One of the main objectives of the Bitcoin blockchain is to eliminate the intermediaries involved in the transaction. Thus, this elimination allows the reduction of the costs of a transaction and allows in particular the "cheap" payments at the international level. According to a report 'The Fintech 2.0 paper: rebooting financial services', written in collaboration between Santander, InnoVentures, Oliver Wyman, and anthemis group, the savings due to the use of blockchain technology could reduce the costs related to international payments, security, and compliance by 15–20 billion in 2022. This is therefore a major challenge for players in the banking sector (Ali et al., 2014).

9.2.2.2.4 *Speed of Transactions*

The characteristic of the speed of transactions constitutes at the same time a lever and a brake to the development of the blockchain. Indeed, transactions can be of several types.

Here we will note two types of transactions: international transactions, and transactions carried out on the financial markets (Özyilmaz et al., 2018). Where international transactions can be carried out in several days, other transactions (financial markets) can be ordered in a few microseconds. Thus, the characteristics of the blockchain do not allow for very high transaction speeds. For example, the Bitcoin blockchain can only generate a block, and therefore the transactions it contains, every 10 minutes, given the Proof of Work method adopted. As a result, we may conclude that high-frequency trading on the blockchain is now not possible. However, when it comes to international exchanges, transactions on a blockchain can be carried out much faster than what banks currently offer. This characteristic can therefore be qualified according to the type of transaction (Wu et al., 2021).

9.3 Conclusion

Bitcoin has become popular, it is important to start asking questions about its use in illegal transactions because of the anonymity it gives to its users. Criminal activity related to Bitcoin does exist, but it is rare. It is not the perfect currency for crime because it is easily traceable when the number of transactions exceeds the norm. It should also be noted that companies offering exchange platforms must, know their customers and therefore perform KYC (Know Your Customer) procedures. It is possible that Bitcoin could be used as a means of tax evasion, and today buying Bitcoin could be like buying gold bullion and storing it without having to declare it to the tax authorities, since the virtual currency poses no storage problems and remains unidentifiable thanks to its anonymity. Because of their particularities (extraterritoriality and absence of regulatory body) and their mode of operation, 5XAU is a code, which means: the code for the quotation of a troy ounce of gold on the financial markets, according to the ISO 4217 standard 6 Forex is the abbreviation of Foreign and Exchange which designates the world currency market, that is to say, the market on which currencies from all over the world are traded (Macedo, 2018). In terms of overall volume, FOREX is the second largest financial market in the world after the interest rate market. KYC is the name given to the process of verifying the identity of a company's customers. The term is also used to refer to the banking regulations that govern these activities. Moreover, from a legal point of view, there is a growing interest in finding a legal definition of Bitcoin in many countries. The interest in a definition and the attribution of a legal framework demonstrates certain confidence and the will to control the sustainability of this very special currency (Wu and Tran, 2018). The financial world is interested in this financial novelty, observing its diversification and performance.

Thanks to blockchain technology the actors of supply chain can benefit from a secure and guaranteed information to carry out their transactions. The application of blockchains in cryptocurrency not only reduces risk of information but also eliminates several processing and transaction fees. This allows governments with volatile currencies to have a more stable currency. In this chapter, we studied the link between blockchain. We also discussed the technological evolution of the currency due to blockchain. We showed that blockchains can offer opportunities to change the centralized and inter-mediated practices of financial markets. However, this technology will be accompanied by the emergence of risks that the researchers need to evaluate.

References

Ali, Robleh, Barrdear, John, Clews, Roger, et al. Innovations in payment technologies and the emergence of digital currencies. *Bank of England Quarterly Bulletin*, 2014, 1, pp. 262–275.

Allen, Darcy W. E., Berg, Chris, Lane, Aaron M., et al. *The economics of crypto-democracy*, 2017, Available at SSRN 2973050.

Ametrano, Ferdinando M. *Hayek money: The cryptocurrency price stability solution*, 2016. Available at SSRN 2425270.

Andreev, R. A., Andreeva, P., Krotov, L., et al. Review of blockchain technology: Types of blockchain and their application. *Intellekt. Sist. Proizv*, 2018, vol. 16, no. 1, pp. 11–14.

Andriole, Stephen J. Blockchain, cryptocurrency, and cybersecurity. *IT Professional*, 2020, vol. 22, no. 1, pp. 13–16.

Back, Adam, et al. *Hashcash – A denial of service counter-measure*, 2002.

Brito, Jerry. *Beyond Silk Road: Potential risks, threats, and promises of virtual currencies. Hearing before the Committee on homeland security and governmental affairs United States Senate*, 2013.

Castelluccio, Michael. Getting started with cryptocurrencies. *Strategic Finance*, 2018, vol. 100, no. 1, pp. 55–57.

Cheung, Adrian, Roca, Eduardo, et Su, Jen-Je. Crypto-currency bubbles: An application of the Phillips–Shi–Yu (2013) methodology on Mt. Gox bitcoin prices. *Applied Economics*, 2015, vol. 47, no. 23, pp. 2348–2358.

Davidson, Laura et Block, Walter E. Bitcoin, the regression theorem, and the emergence of a new medium of exchange. *Quarterly Journal of Austrian Economics*, 2015, vol. 18, no. 3, p. 311.

Gao, Yanni, Li, Zhangmi, Taghipour, Atour, et Kang, Dae Seok. Supply chain coordination: A review. *Journal of Advanced Management Science*, 2018, vol. 6, no. 4, pp. 213–2017.

Hassani, Hossein, Huang, Xu, et Silva, Emmanuel. Big-crypto: Big data, blockchain and cryptocurrency. *Big Data and Cognitive Computing*, 2018, vol. 2, no. 4, p. 34.

Heeks, Richard. Understanding" gold farming" and real-money trading as the intersection of real and virtual economies. *Journal for Virtual Worlds Research*, 2009, vol. 2, no. 4, pp. 1–27.

Herbert, Jeff et Litchfield, Alan. A novel method for decentralised peer-to-peer software license validation using cryptocurrency blockchain technology. In: *Proceedings of the 38th Australasian Computer Science Conference (ACSC 2015)*, 2015, Sydney, Australia, p. 30.

Lebosse, Solen, Taghipour, Atour, et Canel-Depitre, Beatrice. Quick response to fluctuations in supply chains: A review. *Journal of Advanced Management Science*, 2017, vol. 5, no. 5.

Macedo, Leonardo. Blockchain for trade facilitation: Ethereum, eWTP, COs and regulatory issues. *World Customs Journal*, 2018, vol. 12, no. 2, pp. 87–94.

Mannaro, Katiuscia, Pinna, Andrea, et Marchesi, Michele. Crypto-trading: Blockchain-oriented energy market. In: *2017 AEIT International Annual Conference*. IEEE, 2017, pp. 1–5.

Mbiatem, Jane, Taghipour, Atour, et Canel-Depitre, Beatrice. Supplier selection approaches for decision makers. In: *Proceedings of the 8th International Conference on Information Communication and Management*, 2018, pp. 108–112.

Miraz, Mahdi H. et Ali, Maaruf. Applications of blockchain technology beyond cryptocurrency. *arXiv preprint arXiv:1801.03528*, 2018.

Mougayar, William. *The business blockchain: Promise, practice, and application of the next Internet technology*. John Wiley & Sons, 2016.

Özyilmaz, Kazim Rifat, Doğan, Mehmet, et Yurdakul, Arda. IDMoB: IoT data marketplace on blockchain. In: *2018 Crypto Valley Conference on Blockchain Technology (CVCBT)*. IEEE, 2018, pp. 11–19.

Pustišek, Matevž et Kos, Andrej. Approaches to front-end IoT application development for the ethereum blockchain. *Procedia Computer Science*, 2018, vol. 129, pp. 410–419.

Ren, Ziyang, Taghipour, Atour, et Canel-Depitre, Béatrice. Information sharing in supply chain under uncertainty. In: *2016 6th International Conference on Information Communication and Management (ICICM)*. IEEE, 2016, pp. 67–71.

Risius, Marten et Spohrer, Kai. A blockchain research framework. *Business & Information Systems Engineering*, 2017, vol. 59, no. 6, pp. 385–409.

Szabo, Nick. Formalizing and securing relationships on public networks. *First Monday*, 1997.

Taghipour, Atour. Evaluation de la collaboration dans une chaîne d'approvisionnement. *Revue française de gestion industrielle*, 2009, vol. 28, no. 1, pp. 26–42.

Taghipour, Atour, Khazaei, Moein, Azar, Adel, et al. Creating shared value and strategic corporate social responsibility through outsourcing within supply chain management. *Sustainability*, 2022a, vol. 14, no. 4, p. 1940.

Taghipour, Atour, Rouyendegh, Babak Daneshvar, Ünal, Aylın, et al. Selection of suppliers for speech recognition products in IT projects by combining techniques with an integrated fuzzy MCDM. *Sustainability*, 2022b, vol. 14, no. 3, p. 1777.

Tapscott, Don et Tapscott, Alex. *Blockchain revolution: How the technology behind bitcoin is changing money, business, and the world*. Penguin, 2016.

Tliche, Youssef, Taghipour, Atour, et Canel-Depitre, Béatrice. Downstream Demand Inference in decentralized supply chains. *European Journal of Operational Research*, 2019, vol. 274, no. 1, pp. 65–77.

Toyoda, Kentaroh, Mathiopoulos, P. Takis, Sasase, Iwao, et al. A novel blockchain-based product ownership management system (POMS) for anti-counterfeits in the post supply chain. *IEEE Access*, 2017, vol. 5, pp. 17465–17477.

Von Haller Gronbaek, Martin. Blockchain 2.0, smart contracts and challenges. *Comput. Law, SCL Mag*, 2016, vol. 1, pp. 1–5.

Wu, Jiajing, Liu, Jieli, Zhao, Yijing, et al. Analysis of cryptocurrency transactions from a network perspective: An overview. *Journal of Network and Computer Applications*, 2021, vol. 190, p. 103139.

Wu, Jiani et Tran, Nguyen Khoi. Application of blockchain technology in sustainable energy systems: An overview. *Sustainability*, 2018, vol. 10, no. 9, p. 3067.

Yen, Ju-Chun et Wang, Tawei. Stock price relevance of voluntary disclosures about blockchain technology and cryptocurrencies. *International Journal of Accounting Information Systems*, 2021, vol. 40, p. 100499.

10

Blockchain Technology: A Potential to Transform Healthcare

Seema Sahai and Sharad Khattar

CONTENTS

10.1 Introduction

Blockchain technology became popular only when the cryptocurrency, Bitcoin, came into existence. The research on Blockchain is still in its infant state (Xu et al., 2019). Blockchain works on the concept of a distributed ledger of all transactions or any event that has been executed and has been shared amongst the people who are part of a particular network. Each transaction that is stored in a Blockchain is verified by all the participants. If not all, a majority of the participants, so when we talk about Blockchain, we take it synonymous with Bitcoin or any other cryptocurrency. However, that is only one field or one area where Blockchain is used. Blockchain technology as such is based on a peer-to-peer network.

There are various kinds of Blockchain networks: one can be a private network and the other can be a public network. In a public network, anybody can join the Blockchain network (Xu et al., 2019). This is how a Bitcoin functions.

DOI: 10.1201/9781003177432-14

In a private network, only the set of people who are authorized can be part of that network.

The technology on which Blockchain functions is based on open ledgers. These ledgers are visible and available to everybody in the network. The reason for its name being Blockchain is because each transaction is stored in a block. Every block has a block header which consists of the hash value of the previous block the data which it's supposed to contain and its own hash value. The first block is termed as the genesis block which is basically the block which was created to start a network. Accordingly, we can say Satoshi Nakamoto was the creator of the Bitcoin block. The concept of Blockchain is actually based on the fact that manipulating or changing a particular block which has certain data, is very difficult. To change any data requires a concept which is known as proof of work (POW). This POW means more than 50% of the people or the nodes which are part of the Blockchain have to agree to this change, secondly, the change or the addition of a block has a time limit and that time limit varies depending on the application. In Bitcoin, it is about 10 minutes. There is a concept of consensus by which modification of blocks can take place (Kuhle et al., 2021).

If we look at the healthcare system, Blockchain can help in maintaining the supply chain of drugs or medication. Any patient can choose a network of his or her doctors, physicians, and pharmacists who are going to supply the medicines and other people who may be involved in the persons' health care. These chain of blocks would be very difficult to alter because they need the consensus of majority of the people in the network so let's say any member of a block has a mala fide intention of changing any detail that has been provided let's say a drug has been provided which has expired and the pharmacist wants to change it but because it is a stored in a Blockchain it is very difficult to change the block. The pharmacist will need the consensus of the other people in the network because all the people in the network have access to the details of the Blockchain which works on an open ledger system which means every node or every person who is connected to the network has details of all transactions on their system. This is the concept of Blockchain to maintain confidentiality and security of the data.

Blockchains work on hash values so each block has a particular hash value. Modifying anything in a block would lead to a change in the hash value now as discussed earlier the blocks are connected to each other forming a chain by using the previous hash value or that is the hash value of the previous block if any block changes the hash value all the subsequent blocks would need to be modified and this would need a lot of computing and again it would need consensus from the people in the network and this is the reason why Blockchain is considered to be a very secure system

A patient's record can be maintained using the Blockchain technology. One concept that can be used as a simple example is the monitoring of the daily sugar level of a diabetic patient. This data can be transferred to a network through a wearable device. Today we use artificial intelligence and IoT for monitoring the health of people we have wearable devices that monitor the health of people and they can directly send this data to a Blockchain network consisting of physicians and pharmacists. Therefore, the record is maintained about daily fluctuations of the patients' vital statistics, and these blocks can be accessed by authorized people, and they can work on using that data and administering help or diagnosing the patient at the right time. Internet of Things is another technology that works with Blockchain and which allows data to get transferred to a Blockchain network. IOT itself helps in raising an alarm when any of the data has crossed certain limits so in case that alarm has been set up at the doctor's end the doctor can go back to the detailed record and see where the changes are taking place and what the reason could be. This is one way of using Blockchain technology for monitoring the data of a patient (Chelladurai et al., 2021).

The objectives of this paper are as follows:

i. The first objective is to see how Blockchain can transform the traditional healthcare system by showing the model of patient data management

ii. The second objective is to determine how Blockchain technology can speed up the process of treatment of a patient. This can also be explained by the patient data management system

iii. The third objective is to give an insight into how crucial data can be stored safely using Blockchain technology and cannot be tampered with. This is explained by the Drug Traceability model explained in this paper

iv. The fourth objective is to determine the challenges that are there in implementing Blockchain in healthcare systems

v. The fifth objective is to understand the future of Blockchain technology in healthcare systems by citing certain use cases.

10.2 Blockchain vs. Traditional Healthcare System

Blockchain works on the concept of peer-to-peer network it is a decentralised system that further works on distributed ledgers which means any transaction is copied onto all the nodes and their ledgers. All the nodes work on a concept known as consensus where they agreed to the fact that all the transactions will be visible to all of them and that if any new transaction occurs one of them is going to verify it. Any change in previous records can be detected by all the members of the network and such change is usually not allowed. The entire process of changing any old transaction is very cumbersome and requires the consensus of a majority of the notes in the network.

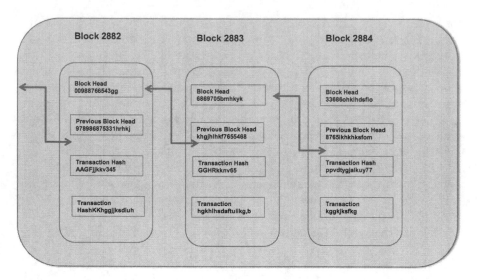

FIGURE 10.1
Structure of a Blockchain.

Blockchain works on the concept of blocks of hash of nodes and of proof of work. These concepts make the Blockchain network immutable or one can say that non-modifiable details of each block can neither be modified nor deleted. Figure 10.1 shows a simple Blockchain and its structure.

There are various types of blockchains. The most popular way of distinguishing types of blockchains are Public Blockchain, Private Blockchain, and Federated Blockchain.

10.2.1 Public Blockchain

These types of blockchains work on an open network. Anybody is allowed to join that network. Once a node is a member of the network that node has access to all the data that is available within the network. Since it is a peer-to-peer network each node has equal rights. This kind of network is seen in cryptocurrencies, especially the Bitcoin. An important aspect of this network is the proof of work or the verification of a block.

10.2.2 Private Blockchain

A Private Blockchain is particular to a set of people who have decided to keep all the data that records the transactions within themselves. It is also a network that works on peer-to-peer but unlike a public Blockchain, not everybody can join the network. Only a few designated people have agreed on using that network as a part of the network.

10.2.3 Federated Blockchain

A federated Blockchain is a Blockchain network that consists of a group of people of different organizations. These groups of people share their ideas their data amongst themselves the reason for this could be improvement in performance or if an organization wants to have healthy competition this sharing of ideas and concepts is a very good platform

Besides these kinds of Blockchain technologies, there are also permission-less and permission-based Blockchain network (Helliar et al., 2020). These concepts are based on who the members of the Blockchain networks are, followed by definitions of hybrid Blockchain.

10.3 Blockchain in Healthcare

The need for Blockchain in the health sector has become particularly important during the time of pandemic where mobile health applications and electronic medical records have been found to be more useful. Moreover, data can be sent to medical personnel rapidly, and data can be accessed very easily at any point. This is the reason why Blockchain becomes very important. The problems of malware, hackers, etc. are avoided with the use of Blockchain (Boireau, 2018).

It can help in identifying the trail of drugs or medications which are provided by pharmaceutical companies with a lot of transparency. All the inputs that go into making that particular drug or medicine can be easily traced, and this certainly helps in the diagnosis and medication of the patients.

Blockchain is also very useful while claiming health insurance since all the data is secured and stored in a particular place. Health insurance claims can be faster and easier.

In times of pandemic, it has been seen that Blockchain has helped in real-time disease reporting and studying the patterns that help identify the origin of the disease and the transmission parameters.

What is needed in healthcare is a network infrastructure, which is secured at all levels. The need for verification and authentication is also one important aspect, which is the need of the hour in healthcare. Blockchain can provide a base where it will give a uniform standard or a uniform access and authorization to all electronic health information. It can be said here that public Blockchain cannot be used in healthcare services because the data that is distributed, needs to be within a very closed private circle. So, usually private Blockchain is used in healthcare systems globally. Another important aspect is to keep the Blockchain technology secure and protected from hackers. This is because information security in health care systems is very important and attacks are very frequent (Tanwar et al., 2020).

While using Blockchain technology it should be kept in mind that data is immutable in Blockchain and therefore whatever data keeps changing very fast and frequently should be avoided and should be kept out of the chain.

The need to have Blockchain technology in the healthcare system over the traditional method of keeping records is mostly because of the decentralized management and immutable databases. Blockchain also helps in tracing data and keeping data away from unauthorized users.

10.4 Applications in Healthcare

There is a wide range of applications that are being used in healthcare based on Blockchain technology. The concept of a distributed ledger in Blockchain helps in the secure transfer of patient medical records, and it also helps in the supply chain of medicine. Many countries around the world have started using Blockchain for enhancing the healthcare system. Estonia began using Blockchain technology as early as 2012 just to secure healthcare data and process transactions. Today in the entire country, healthcare billing is handled by Blockchain. The country also has 95% of health information on this distributed ledger system. Also, 99% of all prescriptions and the information related to it are digital.

The decentralization of record data in Blockchain technology issues that no patient data is ever lost it is immutable and can be retrieved and updated only by blocks involved in the chain. At times data loss causes a lot of distress for health professionals and patients alike. The chronological history of the patient and the monetary transactions that a patient may have done are available for analysis of a patient's condition. The security of Blockchain technology prevents changes from being made either by a hacker or by anyone in the network and therefore the data remains in a very secure and transparent way.

In India, both patients and health care givers benefit from this technology. It helps in removing the complicated process of storing a huge amount of data, and through this technology, all intermediaries are removed from the network. It will also help in expediting insurance claim by any of the patients as it will remove problems like discrepancies in medical records, data manipulation, and so on, as is otherwise seen in the healthcare sector.

In India, the traditional healthcare system which was based on a manual volume-based care system is being replaced by a new value-based care system. Here it is to be seen that the traditional model in India was focused on charging the patients based on the services provided, no matter whether it was relevant or significant, medically. The new value-based healthcare would focus more on customized patient-centric services. This would lead to affordable treatment and superior healthcare services based on big data. Big data is very centric to industry 4.0 and also the new healthcare system which includes things like drug testing, patient data management, Remote patient monitoring clinical trials, and so on.

The entire setup, however, involves a lot of expenses including that of storing and processing of data. India currently spends less than 4% of its GDP on healthcare (World Bank, 2018). A lot more is needed to implement such a large-scale value-based system.

10.4.1 Patient Data Management

The most demanding area in the healthcare system where Blockchain technology could be useful is that of keeping important medical data safe and secure. The biggest worry of all nations is to ensure security in the healthcare system. By using just the digital technology, there have been data breaches of more than 170 million patients between the years 2009 and 2017 in the entire world. Details of credit card and banking information also have been stolen, along with the genomic testing records.

Blockchain ensures an incorruptible decentralized and transparent log of all patient data. Its private Blockchain feature is able to protect sensitive medical data. Besides the security, the decentralized feature of Blockchain allows one platform for patients' doctors and other health care providers with information, which is accurate, quick, and safe.

There are a number of Blockchain healthcare companies that have come up globally and have been working and improving their Blockchain technology for providing people with the desired output some of these companies are Akiri BurstIQ Fatcom Medicalchain Simplyvital Health Embleema and Chronicled to name a few.

The patient data management model has been explained using Figure 10.2.

The patient data management system works on a system where a patient agrees to share his or her data and has something known as a smart contract in a Blockchain. This Blockchain is of course a private Blockchain where only

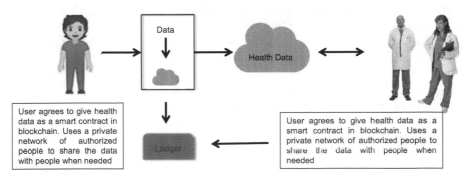

FIGURE 10.2
Model for patient data management

authorized people can share and access data. This data is stored in the cloud and is available to all the people in the network. This may include physicians pharmaceutical pharmacists and other related individuals. Whenever the patient needs the assistance of any of the people in the network the data will be available almost instantly. The patient can be assured of prompt and proper treatment and does not need to worry about purchasing the drugs. Another way of storing data is to store them in smart contracts like a wallet or health wallet. This health wallet may be accessible to various people who need the data for research and other activities. These researchers can use the health data by making a transaction equivalent to that mentioned in the smart contract. Once the transfer of health coin has taken place the data will be released to them. Here no intermediary is involved. The health coin can be used by patients to pay for medical services.

The Blockchain-based patient data management system enables the organization to simplify claim processing, it enables secure medical records and to monitor the pharma supply chain and also collaborate with other network stakeholders (Binaifer & Lakshman, 2017).

Smart contracts can be defined as small computer programs stored in Blockchain which run only when certain preconditions are met. It is used for automating an agreement amongst all the members of a network. So when the required condition exists the transaction takes place. In terms of the data of the patients which may be accessed by researcher, the value of the data is determined according to its size. The researcher who wants to access the data has to first make the payment and in return get the data.

Blockchain ensures an incorruptible, decentralized, and transparent log of all patient data. Its private Blockchain feature is able to protect sensitive medical data. Besides the security, the decentralized feature of Blockchain allows one platform for patients, doctors, and other health care providers with information which is accurate quick and safe.

10.4.2 Drug Traceability and Medical Supply Chain

When we talk about medicine as a layman, how much do we clearly know about it? Is there any surety that the medicine that we have, have not been tampered with? Another question that arises is that is the medicine genuine or is it a fake medicine? Is it coming from a legitimate supplier? These are some of the major concerns of the medical supply chain and drug traceability.

In many countries pharmaceutical supply chain has been impacted by Blockchain technology Blockchain has guaranteed full transparency in the shipping process from the time that a drug is created and a ledger will be formed which will record every step of the way, till it reaches its consumer. Even the labour cost and waste can be monitored through this technology.

Figure 10.3 shows how a model for Drug traceability can work

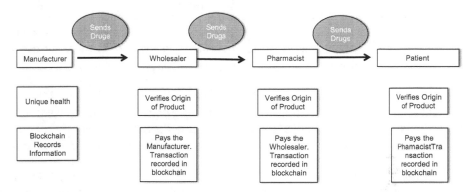

FIGURE 10.3
Drug traceability model

Traceability can be defined as recording of all movements of a product throughout the supply chain. These kinds of applications have been used in many fields.

The infrastructure needed for enabling traceability of a product. Traditionally, a high-end server, a very fast network, and the ability to share data with stakeholders are required. Blockchain removed all these challenges as it removes the need for servers, and there is no authority but there is a lot of transparency.

When the drug manufacturer manufactures a new drug, a new code is assigned to it. The details are stored in a Blockchain and hash is produced. Further, the manufacturer sells the drug to a wholesaler. This wholesaler verifies the origin of the drug and if found correct, does the payment for the transaction in an agreeable method. The wholesaler then approaches a pharmacist. The pharmacist verifies the origin of the drug and then makes the payment for it. The pharmacist then sells the drug to the patient who can verify the origin of the drug and who pays the pharmacist for the drug and again this transaction is recorded in the Blockchain. This way the entire transaction is stored in the Blockchain and can be accessed by all in the network at all times.

10.4.3 Challenges of Blockchain

Blockchain being a new technology has a lot of challenges with it. One major challenge is that healthcare applications have certain unique requirements, that need to be focused on while designing the whole system. Experts are needed to design the system and work on a trial and error-based mechanism. Since development in this technology is still at a very nascent stage, the ongoing implementation of the technology will keep on having certain teething problems for some time (McGhin et al., 2019).

The first challenge is to a certain that there is a need for using Blockchain technology in healthcare to overcome the issues of the traditional method. This needs a lot of convincing from the policymakers, the general citizens who use this system, and the medical experts who would be involved in the network.

In the ledger system, there would always be a lot of people involved in the network. If people are not willing to trust third-party it becomes very difficult to store data in a Blockchain. However, the Blockchain technology has a feature of a permissioned network, and this restricts access to people in a network.

One major challenge of Blockchain technology is the initial cost. Miners use systems that have high computational capability and these systems are expensive. The servers which enable the network to keep functioning require a lot of energy or what we call power and this again is very costly. The energy consumption is very high due to the proof of work and the consensus method which at times can be very wasteful. It has been seen that ever since cryptocurrency involved many people the global energy consumption has been very high.

Another challenge is the security issue. Though Blockchain is considered to be a very secure system because of proof of work and consensus yet its decentralized nature does not assure the blocking of hackers. If a hacker has a high computational network, the Blockchain can be compromised.

Privacy is also sometimes compromised on public Blockchain, especially in the field of healthcare systems. People are reluctant to use the Blockchain technology because they feel that the data may be compromised. Free access to all in the network may sometimes not be a very good idea.

Another big challenge is the public perception of Blockchain technology. People do not believe that this technology will last for long. This is mostly because of the high power consumption and energy consumption.

The legal system of many countries has not yet evolved to an extent that they would support this kind of technology. The policymakers are yet in a quandary as to how to frame the legal system.

10.4.4 Use Cases in India

In India, there is participation from both private and public sectors to develop a healthcare system based on Blockchain technology. The main aim is to enable the health records of patients to map their health information system using their respective unique citizen ID which is stored on the Blockchain platform. The aim is to give an uninterrupted information exchange to all stakeholders in India.

The Blockchain technology has been used to identify five major categories of use cases in the healthcare system in India (PwC India, 2019). Each one of them is briefly outlined in this paper.

The first use case is that of **health profiling**. Here, a personalized health profile of each citizen would be recorded this would be helpful for a centralized mapping of all citizens. This also would be useful in determining rare diseases, rare blood types and facilitate the need for special care and medication. Here, no medical practitioner would be able to access that data without the permission of the citizen. This also means nobody will be able to change or delete the data. Hereby, security will be maintained using the features of Blockchain technology. Another thing would be that all government-accredited hospitals and laboratories would be integrated with the Blockchain and would be allowed to add new data about the patient to the Blockchain of the citizen. Authentication of the user would be very essential. Only on proper authentication will any healthcare personnel be allowed to join the network and the user will be notified about it.

The second use case is of **Citizen registration**. This covers all the registration facilities at one source which means birth and death registration of the citizen and it will serve as a single source for all required data. This would enable the government to communicate with the citizens regarding certain schemes related to a particular age group. This facility can also be integrated with various other schemes like blood donated by the citizen, organ donated by the citizen, etc.

The third use case is that of **health insurance management**. The entire patient data management system can enable a citizen to manage all claims and cashless benefit that may be used by the citizen. This system will be so transparent that there would be no need for any intermediary nor would there be any discrepancy in the claims. Since Blockchain technology records all transactions for good and does not allow any manipulation of data, the data can be referred to at any time. By this technology, false claims can also be eliminated. By bringing about this kind of transparency the system will help the entire nation to move towards equity and avoid any kind of grievances.

The fourth use case would be that of **medication tracking**. This would be linked to drug traceability. This system would contain the entire drug inventory of the nation. It would help to prevent any kind of counterfeit medicine and would also prevent black marketers. In this network, the entire pharma supply chain would be linked to all pharmaceuticals, pharmacies, hospitals, clinics, laboratories, etc. The price of each medicine would also become standardized by using this system. Here we see that a lot of discrepancies in the traditional system can be overcome by using Blockchain technology for tracking.

The fifth use case is **provider care**. By this system Blockchain technology would store all patient data in a single repository. The citizen here would benefit from easy retrieval of data from any place at any time whenever needed. If a citizen happens to travel and a need for a kind of health service arises, the citizen can find out the nearest health service provider from this network. In this system, the feedback of the customers through the government will also be recorded. This will ensure service quality and helps in improving the software and the facilities provided.

10.5 Conclusion

Blockchain has come up to revolutionize healthcare. With Blockchain, patients can be the focal point of all operations and with better security privacy and accessibility. It is not only a highly secure system but it is an open system that can be applied to the medical industry in a number of ways. It helps in reduction of cost and enables secure access to data. Blockchain can be used to create a platform for rebate negotiation and also for contract management. The entire process of contract negotiations will become faster and easier. Blockchain networks allow transparency along with privacy, helping people conceal sensitive patient data and also allowing access and sharing when needed. Hacking a Blockchain would mean that the hackers would require very high computational capabilities again hacking itself is a difficult process because of the distributed ledger system. This leads to the minimization of the attempts of hacking the healthcare system.

Analyzing all these advantages one can say that the healthcare system would improve growth globally if Blockchain were to be applied in all countries. Patient data management is an important task for the well-being of the citizens of any country. Drug traceability will eliminate a lot of malpractices in the pharmaceutical sector. This way corruption and wrong practices will also decrease.

10.6 Limitations and Future Scope

The model can be explored further in cases where it has actually been implemented. This would give an insight into drawbacks of the system which haven't been considered in the study.

With the advance in technology at a great speed, there have been cases of hacking systems using Blockchain technology. This aspect can be dealt with by keeping cybersecurity in mind

References

Binaifer, K., & Lakshman, S. (2017). Blockchain technology in India: Opportunities and challenges. *Deloitte, 17.* https://www2.deloitte.com/content/dam/Deloitte/in/

Documents/strategy/in-strategy-innovation-blockchain-technology-india-opportunities-challenges-noexp.pdf

Boireau, O. (2018). Securing the blockchain against hackers. *Network Security, 2018*(1), 8–11. https://doi.org/10.1016/S1353–4858(18)30006–0

Chelladurai, M. U., Pandian, D. S., & Ramasamy, D. K. (2021). A blockchain based patient centric electronic health record storage and integrity management for e-Health systems. *Health Policy and Technology, 10*(4), 100513. https://doi.org/10.1016/J.HLPT.2021.100513

Helliar, C. V., Crawford, L., Rocca, L., Teodori, C., & Veneziani, M. (2020). Permissionless and permissioned blockchain diffusion. *International Journal of Information Management, 54*, 102136. https://doi.org/10.1016/J.IJINFOMGT.2020.102136

Kuhle, P., Arroyo, D., & Schuster, E. (2021). Building A blockchain-based decentralized digital asset management system for commercial aircraft leasing. *Computers in Industry, 126*, 103393. https://doi.org/10.1016/j.compind.2020.103393

McGhin, T., Choo, K. K. R., Liu, C. Z., & He, D. (2019). Blockchain in healthcare applications: Research challenges and opportunities. In *Journal of Network and Computer Applications* (Vol. 135, pp. 62–75). Academic Press. https://doi.org/10.1016/j.jnca.2019.02.027

PwC India. (2019). *Reimagining health information exchange in India ising blockchain. Pricewaterhousecoopers,* April 28. https://www.pwc.in/assets/pdfs/consulting/technology/it-function-transformation/insights/reimagining-health-information-exchange-in-india-using-blockchain.pdf

Tanwar, S., Parekh, K., & Evans, R. (2020). Blockchain-based electronic healthcare record system for healthcare 4.0 applications. *Journal of Information Security and Applications, 50*, 102407. https://doi.org/10.1016/J.JISA.2019.102407

World Bank. (2018). *Current health expenditure (% GDP).* https://data.worldbank.org/indicator/SH.XPD.CHEX.GD.ZS?locations=IN

Xu, M., Chen, X., & Kou, G. (2019). A systematic review of blockchain. *Financial Innovation, 5*(1). https://doi.org/10.1186/S40854–019–0147–Z

11

A QFD Approach to Design a Blockchain-Based Professional Development Module in Education System for Society 5.0

Vasundhara Kaul and Arshia Kaul

CONTENTS

11.1 Introduction

Education institutions to date are much under the control of educational institutions, and they offer all the facilities related to education. The traditional model is very rigid and has set rules and regulations which are followed. It is, therefore, difficult to surpass the bureaucracy of administration and learn about the education system. There are many stakeholders, teachers, students, and administrative departments in the education system. The education system works well if different stakeholders work together as a unit. The rigidity in the system makes interactions in the current system very cumbersome. There are many details, such as achievements, certifications, and admission-related details, which need a lot of interaction openly. In the current system, it is realised that these details sometimes are not easy to access, and therefore, smooth functioning of the system is hampered (Harris and Wihak, 2017; Lundvall and Rasmussen, 2016; Mayombe, 2017; Müller et al., 2015).

To overcome the rigidity of the traditional educational system, one of the latest technologies, blockchains, needs to be developed for the education

DOI: 10.1201/9781003177432-15

systems. As per Tapscott and Tapscott (2017), blockchains can be considered the second generation of the internet. In the first generation, the focus was only on communication and collaboration, which has shifted to the movement and storage of data rather than only value. Thus, the trusted ledgers, which are secure and help store information related to the money accounts, transcripts, and certificates, are considered to help higher education and provide alternative models for continuous lifelong learning.

The blockchain-based university would be a complete system that provides the infrastructure so that students can create their pathway of learning. It is further stated in Tapscott and Tapscott (2017) that the founder of the Institute of Blockchain Studies has used blockchains to manage the Massively Open Online Courses (MOOCs). Blockchain technology can be used to verify the student who has enrolled for the courses and mastered the online content. The other good characteristic of the technology is that the transactions for payments can be set up using smart contracts for ensuring lifelong plans. Through the technology, both current and future employees would be able to see the return on investment to develop the skills of the individuals in an organisation.

Further, there could be an interaction between the faculty and students, and collaboration is possible in the system. Many universities and institutions have tested blockchain technology. MIT's Media Lab is one of them. The experiment is based on blockchain-based certificates. There are a couple of observations from the experiments that have been previously applied (i) The technology is not a one-stop solution for erroneous certificates and claims of qualifications, but it is a step to reduce the frauds that can be committed of this nature (Nazare et al., 2016); (ii) there is difficulty in managing authentication keys for the issuer and recipient. The use of bitcoins as cash is a better alternative; (iii) giving control of access to the reviewing credentials poses a challenge for security concerns.

Over the years, there have been many changes that have taken place in the technology used in the education space, and digital credentials are becoming popular amongst employers. It is believed that apart from degrees, there are other methods that could be used to showcase the qualifications. There is a continuous collaboration that is taking place between MIT's Media Lab and Learning machine for accessing the blockchain-based credentials and add to their digital resumes. The Registrar of Carnegie Mellon Universities also mentions that adoptions of the blockchain and its initial acceptance are limited compared to any other technology that receives initial resistance and then is used extensively by everyone across the board (Schaffhauser, 2017). Many Universities, such as the Southern New Hampshire University, is also one of the universities considered blockchain implementations for higher education.[1] Further, Clark (2015) has highlighted many blockchain

---

[1] www.youtube.com/watch?v=9qJchEhV-Eo

systems uses in higher education. Some of the benefits are: (i) tracking the project-based education opportunities; (ii) shared repository of learning certificates for universities across global alliance; (iii) global assessment database such as Sony Global Education[2]; (iv) details of continuing professional development and repository of corporate learning and apprenticeships.

From the above studies, it can be highlighted that some universities are trying experimentally to implement blockchain-based systems. It can be stated that the blockchain-based systems will improve the capabilities of the system stakeholders, not only in terms of maintaining records but also in professional development. There are some institutions for which we have highlighted specifically, such as MIT and the University of New Hampshire. Yet, there is limited or almost no use of such a blockchain-based system in India. With the learning happening across borders, many changes are being brought about, but not complete implementations for an overall blockchain-based system have been developed in educational institutions. In this paper, therefore, we aim to assess what such an end-to-end blockchain-based system should comprise for better management of educational systems. We aim to understand the institution's needs and understand what it would want for the professional showcasing and ensuring continuous professional development through the blockchain-based system. We utilise the Quality Function Deployment (QFD) methodology to understand the system requirements (whats) and assess them against the how's against the what's. The assessment is done to understand how they would be able to achieve the system capable enough to develop the capacity of stakeholders (students and faculty) in the blockchain-based educational system. The paper tries to address the following questions (i) what are the requirements for the development of a blockchain-based professional development module, (ii) what is the importance of each of the needs in the implementation, (iii) how will these system requirements be achieved, (iv) which of the how's the most important to start the smooth implementation of the blockchain system

11.2 Literature Review

This section aims to understand the related research work which considers blockchain implementation for education systems. We want to address which educational institutions have implemented blockchains and learn from them to develop a more efficient blockchain-based education system. One of the first universities which stored certificates on a Bitcoin blockchain was the

---

[2] www.sony.net/SonyInfo/News/Press/201602/16-0222E/index.html

University of Nicosia[3] (Sharples, et al., 2016). One of the other institutions was the collaboration of MIT Media Lab Learning initiative with Learning Machine (an enterprise vendor). They developed Blockcerts, open-source software for the verification of blockchain-based certificates. These certificates contained the details such as the name of the recipient, name of issuing institution, date of issue, and other relevant information. These Blockcerts were said to be tamper-proof and were verifiable across the blockchain.[4] Further, a pilot study was done in Malta based on the Blockcerts,[5] and the Federation of State Medical Boards in the US also launched a pilot of issuing documents with Blockcerts to the blockchain.[6] Based on Ethereum, the company SAP TrueRec, a secure and trusted digital wallet for storing professional and academic credentials, was set up in July 2017.[7]

Another blockchain project was set up by TNO, Netherlands Organization for Applied Scientific Research. This helped supply information digitally and only shared minimum personal data stored in encrypted form in a wallet of the personal cellphone. The confirmation is provided of a person.[8]

Notarisation is similar to certification, and the ownership, existence, and integrity of the documents are important. The Apostille notarisation service and use cases such as the car ownership and other details such as registration of licences are described by McDonald and Oliverio.[9] Kolvenbach et al. (2018) have also discussed the demo blockchain for education. They discussed the details of Ethereum-based blockchain and developed a model for the same.

From the literature, it can be observed that there are still only pilot projects that have been set up. There are no major full-time projects that have been set up. They are still in progress.

11.2.1 Research Gap and Motivation

From the current literature, it can be observed that blockchain as a concept has only recently come to the fore. Moreover, the use of blockchains in the education sector and education institutions is limited. Not many people are aware of its uses in educational institutions. Since they are not aware of the use, further the application of maintaining the system based on blockchain

---

[3] http://www.educationmalta.org/blockcerts -to-be-developed-in-malta/
[4] Digital Certificates Project, Certificates, Reputation, and the Blockchain – MIT MEDIA LAB.. Retrieved from: http://certificates.media.mit.edu/
[5] Case Study Malta | Learning Machine. Retrieved from https://www.learningmachine.com/case-studies-malta
[6] Case Study FSMB | Learning Machine. Retrieved from https://www.learningmachine.com/case-studies-fsmb
[7] MeetTrueRecbySAP:TrustedDigitalCredentialsPoweredbyBlockchain.Retrievedfromhttps://news.sap.com/meet-truerec-by-sap-trusted-digital-credentials-powered-by-blockchain/
[8] https://www.techruption.org/usecase/xxcvxcvxcv/
[9] Apostille White Paper. Retrieved from https://nem.io/wp-content/themes/nem/files/ApostilleWhitePaper.pdf

is not seen. Thus, the motivation of the research is to determine how a blockchain-based system be set up in any educational institution.

11.3 Methodology

The paper's focus is to develop an efficient blockchain-based education system. For this, we need to understand the components required for the system development. The system can only be developed if we understand what the outcome must be. An education system needs to keep developing dynamically to help deliver up-to-date knowledge for the future workforce. For the development of the educational system, each stakeholder needs to keep developing themselves, may it be the faculty, students, or any other allied individuals with the institution. This chapter aims to develop such a dynamically developing blockchain-based system.

For this, the methodology is used in the Quality Deployment Function (QFD). This methodology helps to understand what is required as a deliverable of a blockchain-based system. These are highlighted as the "What's" which suggests the customers' voice. In this methodology, the characteristics of the product or service will help to make the outcome of high quality in the customer's minds. It is simply the planning tool used in research to identify the Voice of Customers (VoC) or, in other words, 'what' the customers need (Hwarng and Teo, 2001).

In this case, the voice of the customers is the system stakeholders, such as the teachers, students, heads of institutions, and directly or indirectly parents. Once the needs are identified, it is necessary to find out which has the highest priority level, to ensure that the customers are satisfied by achieving those needs first. It is also necessary to understand how the needs will be achieved and group them under the "how's" of the system. The outcome of the final evaluation of the system through the QFD is the assessment of which is the most important method one must concentrate on to achieve the needs for satisfying the needs of customers. It helps transfer the understanding of the "what's" to decide design parameters for the product or service to achieve customer satisfaction (Eldin, 2002). The House of Quality (HoQ) is the easy-to-understand representation of QFD to understand the "what's" of the customers, "how's" to achieve the "what's" and the relationship of between the "How's" and "What's" (Hauser and Clausing, 1998; Chou, 2004). Figure 11.1 highlights the broad steps of QFD.

QFD is a structured approach that highlights that it is extremely important to understand the customer's requirements and it considers empathy towards the customer as the most important (Bouchereau & Rowlands, 2000). It is a methodology used to design a product or service depending on the customer's needs (Prasad and Chakraborty, 2013).

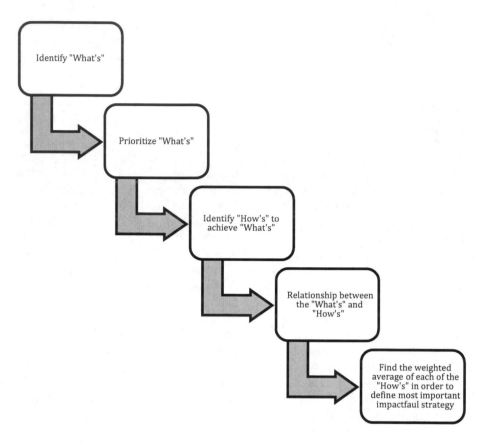

FIGURE 11.1
Methodology of research.

Figure 11.1 gives in detail an overview of the research methodology, including the steps of QFD.

11.4 Case Illustration

The higher educational university under consideration (name undisclosed due to confidentiality reasons) is looking for continuous improvements in its systems so that it can provide the best environment for teaching-learning. The continuous improvements, in general, could be considered in many ways: (i) improvements in terms of use of technology (ii) changes in the syllabus as per industry standards (iii) infrastructural improvements (iv) changes in the international collaboration for teaching-learning (v) international exchange programmes and many new adaptations that institutions keep making in the system so that their institution is considered as one of the trusted institutions in the country and subsequently worldwide. In the case of the current higher education university, the requirement is to develop a system for professional development for the current and future. Professional development could be at two levels, one for the development of the faculty members in order to keep them up-to-date with the latest trends and knowledge base so that they can provide effective learning experiences to their students using the latest pedagogies. On the other hand, professional development could also be for the students to develop their skills for their future jobs. The university wishes to provide the platform for ensuring that the students can get access to add-on professional courses to upgrade their skill sets and make them future job ready.

In such a situation, the system has to be developed so that there is an easily available structure for the professional development of the stakeholders of the system, as discussed. There are many ways to develop such a structure. One of the latest technologies is blockchain technology, which can develop an integrated platform for overall professional development. The questions in front of the university are that even if they know which technology they might want to implement, what should be the output of such a system? What should be the functionality of such a system? Further, once one can establish the needs of the system, how would one achieve the needs of the system? These kinds of situations typically fit the implementation of the QFD model.

In the QFD-based model, as described in the methodology section, we can list down the requirements of the blockchain-based professional development module for the university. These requirements are highlighted as the "what's" of the system. In relation to the "what's", we highlight the "how's" of achieving the university's efficient blockchain-based professional development system. Table 11.1 gives the "what's" (R1–R20), and Table 11.2 gives the "how's" (H1–H8), which are defined based on the discussion with the stakeholders and the literature in the area (Mikroyannidis et al., 2020). These "how's" will define the way in which the efficient blockchain-based system is developed.

The relationship of the what's and how's are determined using the QFD approach. The results and the scales for weights used for the evaluation are discussed in the Results and Discussion section.

11.5 Results and Discussion

As discussed in the methodology section, QFD has been utilised for determining the "what's" and "how's" for achieving an efficient and secure blockchain-based professional development system for the case university under consideration. In this section, we highlight the final results obtained from the evaluation. While completing QFD, we have used the following

TABLE 11.1

Requirements for the Blockchain-Based Professional Development System

R1	Option of issuing certifications	R11	Smart contracts for collaborations
R2	Option of issuing co-branded certifications	R12	Provision for the use of shared databases
R3	MIS tracker for certifications	R13	On-job training for teachers based on the teacher's lifecycle in the institution
R4	Plagiarism checking for research documents	R14	Collaborative teaching and learning
R5	Authentication of certification	R15	Recommendations of new courses based on searching history
R6	Institution's research work tracking	R16	Career assessment for students
R7	Comparative reporting of research work done across various domains	R17	Recommendation of add-on certifications based on career assessment
R8	Provision for teachers to offer asynchronous courses through the blockchain system	R18	Online transactions for course fee collection through blockchain
R9	Provision for teachers to offer synchronous courses through the blockchain system	R19	Placement opportunities and job-search for students passing out from the institution
R10	Assessment through blockchain for certifications	R20	Access to online career counselling

Source: Mikroyannidis et al. (2020), Tapscott and Tapscott (2017), and Nazare et al. (2016).

TABLE 11.2

How's for the Blockchain-Based Professional Development System

How's		How's	
H1	User-friendly interface	H5	Easy scalability
H2	Use of code that is easy to upgrade as required	H6	Enhanced security
H3	Trackable data storage	H7	Privacy and identity protection
H4	Archiving information for up to 10 years	H8	Decentralised control through maker and checker options

scale for evaluating the importance of the "what's" (1 – least important, 5 – most important) and for evaluating the influence of "how's" on each of the "what's" the scale used was (0 – no influence, 1 – low influence, 3 – medium influence, 9 – high influence). Table 11.3 shows the evaluation of "how's" in relation to the "what's".

The "what's" represent the requirements of the university from the blockchain-based system for the purpose of supporting professional development of all stakeholders. The "what's" from the perspective of faculty members were based on the development of their research, delivery of courses and

TABLE 11.3

Evaluation of "How's" in Relation to "Whats" by QFD

		Weights	User-Friendly Interface	Use of Code That Is Easy to Upgrade as Required	Trackable Data Storage	Archiving Information for Up to 10 Years	Easy Scalability	Enhanced Security	Privacy and Identity Protection	Decentralised Control through Maker and Checker Options
R1	Option of issuing certifications	4	9	3	9	3	9	3	0	9
R2	Option of issuing co-branded certifications	4	9	3	9	3	9	3	0	9
R3	MIS tracker for certifications	5	3	9	9	9	3	9	3	3
R4	Plagiarism checking for research documents	4	1	1	0	0	3	0	0	3
R5	Authentication of certification	4	9	3	3	1	9	9	3	9
R6	Institution's research work tracking	4	3	1	3	9	9	1	1	3
R7	Comparative reporting of research work done across various domains	3	1	1	3	9	3	0	0	1
R8	Provision for teachers to offer asynchronous courses through the blockchain system	5	3	9	9	3	9	9	1	3

(Continued)

TABLE 11.3 (Continued)

Evaluation of "How's" in Relation to "Whats" by QFD

	Weights	User-Friendly Interface	Use of Code That Is Easy to Upgrade as Required	Trackable Data Storage	Archiving Information for Up to 10 Years	Easy Scalability	Enhanced Security	Privacy and Identity Protection	Decentralised Control through Maker and Checker Options
R9	Provision for teachers to offer synchronous courses through the blockchain system	5	9	9	3	3	9	9	3	9
R10	Assessment through blockchain for certifications	5	3	3	9	9	9	9	9	9
R11	Smart contracts for collaborations	4	3	1	1	3	1	9	1	3
R12	Provision for the use of shared databases	3	3	1	3	1	3	1	1	1
R13	On-job trainings for teachers based on the teacher's lifecycle in the institution	5	9	9	9	9	9	9	3	9
R14	Collaborative teaching and learning	5	9	9	9	9	9	9	3	3
R15	Recommendations of new courses based on searching history	4	1	3	9	3	3	3	3	3

(Continued)

TABLE 11.3 (*Continued*)

Evaluation of "How's" in Relation to "Whats" by QFD

	Weights	User-Friendly Interface	Use of Code That Is Easy to Upgrade as Required	Trackable Data Storage	Archiving Information for Up to 10 Years	Easy Scalability	Enhanced Security	Privacy and Identity Protection	Decentralised Control through Maker and Checker Options
R16 Career assessment for students	4	9	9	3	1	9	3	9	3
R17 Recommendation of add-on certifications based on career assessment	3	3	9	3	0	3	3	3	3
R18 Online transactions for course fee collection through blockchain	5	9	9	9	9	9	9	9	9
R19 Placement opportunities and job-search for students passing out from the institution	4	3	3	9	3	9	3	3	3
R20 Access to online career counselling	4	3	1	3	0	3	3	3	9
		446	430	508	389	568	475	259	456

courses for learning. In the case of the students, it covered add-on courses for getting them job-ready and help them achieve their career development goals.

Dependent on the final weight average calculations which are used in the QFD methodology, it can be observed the "how's" in decreasing order of total scores are (i) easy scalability with a score of 568, the trackable data storage with a score of 508 and privacy and identity protection with the lowest score of 259. This shows that as of the current status, we have been able to analyse the most essential "how" to achieve the efficiency of a blockchain-based professional development module for the university.

11.6 Conclusion and Future Scope

This research aims to assess what are the requirements to design a blockchain-based professional development system for the university under consideration. The voice of the stakeholders is considered in designing the system. The QFD methodology is used for the assessment of the current system. The implementation of the blockchain-based system will ensure that the university and stakeholders are in a better position not only their objectives but also the plan the process of professional development and growth. As Society 5.0 requires education systems to be built in a manner that focus on wholistic development of individuals; therefore this comprehensive blockchain-based system meets those objectives as it supports all stakeholders and also helps to track data for future upgrades and enhancements. The current study is taken for a particular university as a test case and is generalisable to any other educational institution. The limitation of the study is that this is considered for a specific country; there may be a requirement of changing the design such that we are able to modify it for country-specific requirements.

References

Bouchereau, V., & Rowlands, H. (2000). Methods and techniques to help quality function deployment (QFD). *Benchmarking: An International Journal, 7*(1), 8–20. https://doi.org/10.1108/14635770010314891

Chou, S. M. (2004). Evaluating the service quality of undergraduate nursing education in Taiwan – Using quality function deployment. *Nurse Education Today, 24*(4), 310–318.

Clark D. (2015, September 12). 10 Ways Blockchain could be used in education [Opinions]. In *OEB Insights*. Retrieved from https://oeb-insights.com/10-ways-blockchain-could-be-used-in-education/

Eldin, N. (2002). A promising planning tool: Quality function deployment. *Cost Engineering*, 44(3), 28.

Harris, J., & Wihak, C. (2017) To what extent do discipline, knowledge domain and curriculum affect the feasibility of the recognition of prior learning (RPL) in higher education? *International Journal of Lifelong Education*, 36(6), 696–712, https://doi.org/10.1080/02601370.2017.1379564

Hauser, J. R., & Clausing, D. (1988). *The house of quality*. Harvard Business Review, May-June 1988.

Hwarng, H. B., & Teo, C. (2001). Translating customers' voices into operations requirements – A QFD application in higher education. *International Journal of Quality & Reliability Management*, 2(2), 23–26.

Kolvenbach, S., Ruland, R., Gräther, W., & Prinz, W. (2018). Blockchain 4 education. In *Proceedings of 16th European Conference on Computer-Supported Cooperative Work-Panels, Posters and Demos*. European Society for Socially Embedded Technologies (EUSSET).

Lundvall, B.-Å., & Rasmussen, P. (2016) Challenges for adult skill formation in the globalising learning economy – A European perspective. *International Journal of Lifelong Education*, 35(4), 448–464.

Mayombe, C. (2017) An assessment of non-formal education and training centres' linkages with role-players for adult employment in South Africa. *International Journal of Lifelong Education*, 36(3), 339–358.

Mikroyannidis, A., Third, A., Domingue, J., Bachler, M., & Quick, K. (2020). Blockchain Applications in lifelong learning and the role of the semantic blockchain. In Sharma, R. C., Yildirim, I I., & Kurubacak, G., eds., *Blockchain technology applications in education* (pp. 16–41). IGI Global.

Müller, R., Remdisch, S., Köhler, K., Marr, L., Repo, S., & Yndigegn, C. (2015) Easing access for lifelong learners: A comparison of European models for university lifelong learning. *International Journal of Lifelong Education*, 34(5), 530–550.

Nazare, J., Duffy, K., & Schmidt, J. P. (2016, June 2). *What we learned from designing an academic certificates system on the blockchain*. MIT Media Lab. Retrieved from https://medium.com/mit-media-lab/what-we-learned-from-designingan--academic-certificates-system-on-the-blockchain-34ba5874f196

Prasad, K., & Chakraborty, S. (2013). A quality function deployment-based model for materials selection. *Materials & Design*, 49, 525–535.

Schaffnauser, D. (2017). *Blockchain: Letting student own their credentials*. Campus Technology. Retrieved from https://campustechnology.com/articles/2017/03/23/blockchain-letting-students-own-their-credentials.aspx

Sharples, M. et al. (2016). *Innovating pedagogy 2016: Open university innovation report 5.*

Tapscott, D., & Tapscott, A. (2017). The blockchain revolution and higher education. *Educause Review*, 52(2), 11–24.

Index